Qizhen Humanities and Social Sciences Library

Taking advantage of the interdisciplinary strength of Zhejiang University, "Qizhen Humanities and Social Sciences Library" seeks to build bridges between social science academics in China and abroad. Whether it pertains to the arts or sciences, the past or the present, the east or the west, pure or applied, the subject matter seeks to promote publications that represent academic excellence, cultural quintessence and the research cutting edge of China's higher education.

Weiwen Zhang · Yongping Zhang

Digital Urban Governance in China

Foundations and Applications

INTERNATIONAL PUBLISHING PROGRAM OF ZHEJIANG UNIVERSITY
浙江大学学术精品走出去项目

ZHEJIANG UNIVERSITY PRESS
浙江大学出版社

Springer

Weiwen Zhang
School of Public Affairs
Zhejiang University
Hangzhou, Zhejiang, China

China Institute of Urbanization
Zhejiang University
Hangzhou, Zhejiang, China

Yongping Zhang
School of Public Affairs
Zhejiang University
Hangzhou, Zhejiang, China

China Institute of Urbanization
Zhejiang University
Hangzhou, Zhejiang, China

ISSN 2731-5304 ISSN 2731-5312 (electronic)
Qizhen Humanities and Social Sciences Library
ISBN 978-981-96-6324-8 ISBN 978-981-96-6325-5 (eBook)
https://doi.org/10.1007/978-981-96-6325-5

Jointly published with Zhejiang University Press
The printed edition is not for sale in China Mainland. Customers from China Mainland, please order the print book from Zhejiang University Press.

This book is supported by the Humanities and Social Science Fund of the Ministry of Education of China (23YJC630244), the Zhejiang Provincial Natural Science Foundation of China (LY24D010001), and the Fundamental Research Funds for the Central Universities (Zhejiang University; S20250014).

This Springer imprint is published by the registered company Springer Nature Singapore Pte Ltd.
The registered company address is: 152 Beach Road, #21-01/04 Gateway East, Singapore 189721, Singapore

Preface

Urban governance is a specific manifestation of national governance within a particular geographical and administrative unit. It involves multiple stakeholders, including the government, businesses, social organizations, and residents, working through a specific institutional arrangement to achieve a dynamic process of coordinating public and private interests through cooperation and consultation on public affairs. Urban governance is a crucial part of the modernization of the national governance system and its capacity. Advancing the modernization of urban governance has become an important breakthrough and foundational work in realizing national governance modernization. As urbanization progresses rapidly, urban issues have become more complex and diverse, significantly increasing the difficulty and challenges of urban governance. At the same time, technological advancements have brought new opportunities for improving urban governance, prompting a shift from traditional urban governance to digital urban governance. This transformation involves gradually forming a new governance model that enhances urban governance capabilities through next-generation information and communication technologies such as big data, cloud computing, the Internet of Things (IoT), and artificial intelligence.

This book attempts to understand digital urban governance from two perspectives: technology-driven governance and data-driven governance. From a technology empowerment perspective, the application of next-generation information and communication technologies in urban governance has accelerated. These technologies are integrated across all stages of urban planning, construction, management, and operation, serving multiple stakeholders such as the government, businesses, social organizations, and citizens. This integration innovates urban governance mechanisms and improves governance levels. From a data-driven perspective, the process of urban governance is accompanied by the generation of vast amounts of urban data. Such rich data helps us scientifically understand the characteristics of governance entities (such as governments and businesses) and governance objects (such as buildings and roads) involved in the urban governance process. These massive amounts of real-time data also make human behavior predictable, which in turn enables the provision of targeted policy measures and accurate, efficient, and convenient services.

Chapter 1 introduces the basic concepts, developmental evolution, and theoretical foundations of digital urban governance. Chapters 2 and 4 explore how next-generation information and communication technologies empower urban governance, specifically focusing on typical technologies such as IoT, big data, cloud computing, and artificial intelligence, as well as typical cases like the City Brain and health codes. Chapters 3 and 5 discuss how data drives urban governance, focusing on typical urban big data such as mobile signaling data and smart card swipe data, as well as typical applications in traffic and population governance. Chapter 6 systematically explores the future of digital urban governance, including potential opportunities and challenges.

The book was co-authored by me and several young colleagues, with the following specific contributions: Chap. 1 was written by Zhang Yongping and me; Chaps. 2, 3, and 5 were authored by Zhang Yongping, with Dong Zhaoyingzi contributing to the mobile signaling big data case; Chap. 4 was written by me; Chap. 6 was completed by Xu Yuanshuo. I also handled the overall editing work. I would like to give special thanks to Dr. Zhang Yongping for his significant contribution to the successful completion of this book. I am also grateful to big data expert Wang Xiaoliang for his strong support in completing the mobile signaling big data case, and to Dr. Song Yang for providing valuable suggestions on the theoretical foundation of digital urban governance in Chap. 1. My thanks go to PhD students Jin Han and Liu Zekun for their contributions to the City Brain and health code cases.

Throughout the process of conceptualizing, writing, and publishing this book, we have received enthusiastic assistance and invaluable advice from many leaders, professors, and friends, to whom we express our sincere gratitude. I am particularly grateful to editor Chen Peiyu for her hard work, whose valuable suggestions have greatly enhanced the book. I also appreciate the dedication of research assistant Wu Yan from Zhejiang University's China Institute of Urbanization during the proofreading process. This book is intended as a reference for undergraduate and graduate students in public administration-related fields and is also suitable for professionals working in digital urban governance.

This book is supported by another two research institutes of Zhejiang University, namely, the ZJU-CMZJ Joint Lab on Data Intelligence and Urban Future, and the Lab of Inclusive Smart Governance for Urban-Rural Integration. Besides, this book is funded by the Humanities and Social Science Fund of the Ministry of Education of China (23YJC630244), the Zhejiang Provincial Natural Science Foundation of China (LY24D010001), and the Fundamental Research Funds for the Central Universities (Zhejiang University; S20250014).

Digital urban governance is still a new concept, and there is currently a lack of systematic theories and frameworks for digital urban governance. Moreover, as social and economic development accelerates and information technology continuously evolves, new urban governance issues and challenges will arise. Theories, concepts, connotations, and frameworks related to digital urban governance are also in the process of continuous development and improvement. Therefore, systematically interpreting digital urban governance is a challenging task. Due to the limitations of the authors, this study invites further discussion and more in-depth analysis. We will strive to enrich and improve the content of this book in future editions.

Hangzhou, China
November 2025

Weiwen Zhang

Contents

1 Introduction 1
1 Urban Governance and Digital Urban Governance 1
1.1 Urban Areas 1
1.2 Governance and Urban Governance 2
1.3 Digital Urban Governance 4
2 The Evolution of Digital Urban Governance 5
2.1 E-Government 5
2.2 Digital City 6
2.3 Smart City 7
2.4 Digital Urban Governance 11
3 Theoretical Foundations of Digital Urban Governance 14
3.1 New Public Management Theory 15
3.2 Digital Governance Theory 16
4 Organization of the Book 20
References 22

2 Next-Generation Information and Communication Technology and Digital Urban Governance 25
1 The Internet of Things and Digital Urban Governance 26
1.1 What Is the Internet of Things? 26
1.2 The Application of the Internet of Things in Urban Governance 27
2 Big Data and Digital Urban Governance 28
2.1 What Is Big Data? 28
2.2 Urban Big Data Platforms 29
3 Cloud Computing and Digital Urban Governance 30
3.1 What Is Cloud Computing? 30
3.2 The Application of Cloud Computing in Urban Governance 32
4 Mobile Internet and Digital Urban Governance 33
4.1 What Is Mobile Internet? 33
4.2 The Application of Mobile Internet in Urban Governance 34

5 Artificial Intelligence and Digital Urban Governance 36
5.1 What Is Artificial Intelligence? 36
5.2 The Application of Artificial Intelligence in Urban Governance 37
6 Blockchain and Digital Urban Governance 38
6.1 What Is Blockchain? 38
6.2 The Application of Blockchain in Urban Governance 39
References 40

3 New Types of Urban Data and Digital Urban Governance 43
1 Smart Card Data 44
1.1 Introduction 44
1.2 Research and Application 45
2 Smartphone Data 47
2.1 Introduction 47
2.2 Research and Application 49
3 Points of Interest 50
3.1 Introduction 50
3.2 Research and Application 51
4 Social Media Data 52
4.1 Introduction 52
4.2 Research and Application 53
5 Trajectory Data 53
5.1 Introduction 53
5.2 Research and Application 54
References 56

4 Case Studies on Technology-Driven Urban Governance 59
1 City Brain 59
1.1 Background 59
1.2 Key Approaches 63
1.3 Achievements 70
1.4 Lessons Learned 72
2 Health Code 74
2.1 Background 74
2.2 Main Approaches 76
2.3 Achievements 80
2.4 Lessons Learned 83
References 86

5 Case Studies on Data-Driven Urban Governance 89
1 Identifying Companion Travel Behavior Using Subway Card-Swipe Data 89
2 Using User Cycling Big Data to Support Geofencing Planning 95
3 Using Mobile Signaling Big Data to Identify Individual Movement Trajectories 100

3.1 Assessing the Vitality of Characteristic Towns 100
3.2 Improving Public Transportation Efficiency 106
3.3 Evaluating Policy Implementation Effects 110
References 113

6 The Future of Digital Urban Governance 115
1 Urban Governance in the Age of Artificial Intelligence 115
1.1 From Digital Governance to Smart Governance 115
1.2 Future Prospects of Artificial Intelligence in Urban Governance 120
2 Opportunities Brought by Data Openness to Urban Governance 124
2.1 Significance of Data Openness 125
2.2 Future Prospects for Data Openness 127
3 Challenges Facing Future Urban Governance 132
3.1 Risks of Applying Artificial Intelligence to Urban Governance 132
3.2 Challenges Brought by Data Openness 135
References 137

About the Authors

Weiwen Zhang is a professor at the School of Public Affairs, Zhejiang University, the director of China Institute of Urbanization, and the director of the ZJU-CMZJ Joint Lab on Data Intelligence and Urban Future. She received her Ph.D. degree from Zhejiang University. Her research interests include inter-governmental cooperation issues in urbanization and digital urban governance.

Yongping Zhang is an assistant professor at the School of Public Affairs, Zhejiang University and the deputy director of the ZJU-CMZJ Joint Lab on Data Intelligence and Urban Future. His research interests include urban data analytics, digital city governance, smart cities, sustainable transport, and geographic information science.

Chapter 1
Introduction

1 Urban Governance and Digital Urban Governance

1.1 Urban Areas

Urban areas have gradually developed alongside human civilization. In primitive societies, humans lived in caves, trees, or other communal arrangements without permanent settlements. With the advancement of productivity, human society gradually established fixed settlements with functions such as military defense, religious worship, and commodity exchange, giving rise to cities. Today, the functions of cities have become increasingly complex and refined: the roles of military defense and religious worship have significantly diminished, while residential and occupational functions remain crucial. At the same time, the adequacy of services such as consumption, education, technology, and culture has become key factors in determining a city's prosperity.

The concept of the city is highly diverse and lacks a universally accepted definition due to differences in national systems. Generally speaking, cities are understood through three primary characteristics: population, industries, and functions. Cities have higher population densities and larger populations, serve as settlements dominated by secondary and tertiary industries, and act as hubs of commerce, transportation, and cultural and educational industries. Additionally, they function as political, economic, and cultural centers within a specific regional scope (Wu & Li, 2010).

Different academic disciplines offer varying perspectives on cities. Urban planning focuses on spatial layout and the use of planning and design to chart future development and promote growth. Geography examines natural or human geographical elements, spatial characteristics, and the relationship between human activities and geographical features. Sociology addresses social issues that arise during urban development, such as "urban villages," segregation, and crime. Meanwhile, economics explores the agglomeration characteristics of urban economic elements, focusing on the patterns of various economic activities. Each discipline has its specific

W. Zhang and Y. Zhang, *Digital Urban Governance in China*, Qizhen Humanities and Social Sciences Library, https://doi.org/10.1007/978-981-96-6325-5_1

focus and perspective, collectively forming a differentiated yet interconnected knowledge system that advances our understanding of cities (Yang, 2017).

Since the reform and opening-up, China has undergone rapid urbanization. Considering China's population, economy, and spatial scale, this process is unprecedented in human history. According to data from the National Bureau of Statistics, the urban resident population in 2019 reached 848.43 million. From 1978 to 2019, China's urbanization rate rose from 17.9% to 60.6%, with an average annual increase of about one percentage point. Based on global urbanization trends, China remains in the rapid development stage, where urbanization rates range from 30 to 70% (Northam, 1979). As such, China's rapid urbanization is expected to continue, evolving into a higher-level, people-centered new model of urbanization.

It appears we have entered an era centered on cities. As demonstrated by China's rapid urbanization, an increasing number of people are choosing to move to urban areas. This phenomenon is also evident globally, with people migrating from rural to urban areas and from smaller to larger cities. As a result, global hubs like New York, London, Tokyo, Shanghai, and Singapore have emerged, driving social, economic, and cultural development worldwide. Compared to rural areas, cities offer more and higher-quality opportunities for employment, housing, leisure, education, and healthcare, which are increasingly critical for holistic human development in modern society. Urban living environments can also be enhanced through greening, the promotion of green technologies, and the establishment of efficient transportation networks. This model of urban concentration effectively addresses global climate change and promotes sustainable development.

1.2 Governance and Urban Governance

The word "governance" in English is etymologically related to terms such as "government", "rule", "administration", and "behavioral norms." Therefore, the earliest concept of governance primarily referred to the actions of the government in response to natural or social events (Ye et al., 2021). With the transformation of government functions in Western countries, the rise of privatization, and theories like New Public Management, governance gained new meanings. According to the Commission on Global Governance's report *Our Global Neighborhood*, governance is defined as "the sum of the many ways individuals and institutions, public and private, manage their common affairs. It is a continuing process" (Gao & Que, 2020). Through formal and informal institutional arrangements, diverse actors participate in governance, coordinate conflicts of interest, and ultimately take rational action. Governance has become a crucial issue for China's current and future development. The Third Plenary Session of the 18th Central Committee of the Communist Party of China (CPC) first proposed the strategic goal of "comprehensively deepening reform to improve and develop the socialist system with Chinese characteristics and advancing the modernization of the country's governance system and governance capacity." The Fourth Plenary Session further outlined the requirements for "promoting multi-level and multi-field governance in accordance with the law, emphasizing systemic, lawful, comprehensive, and source governance."

Urban governance can be seen as the concrete manifestation of national governance within a specific geographic and administrative area. It involves diverse actors—including governments, enterprises, social organizations, and residents—collaborating and negotiating on public affairs management and service provision through institutional arrangements to harmonize public and private interests (Huang & Zhang, 2016). Urban governance is a significant component of modernizing the national governance system and capacity. Advancing urban governance modernization has become a fundamental and breakthrough task in achieving national governance modernization (Yang & Zhang, 2020).

Urban governance evolved from urban management. Initially, urban management was conducted solely by governments, handling public affairs and providing public services independently. However, as cities developed rapidly, accompanying urban issues (commonly referred to as "urban diseases") grew increasingly complex and multifaceted. For instance, enhancing sustainable urban development involves various domains such as land use, environment, energy, transportation, and population. It also intersects with improving economic efficiency, promoting urban equity, and fostering civic awareness. In modern society, relying solely on the government to manage cities has become increasingly challenging. Faced with these realities, the concept of governance emerged, marking a transition from urban management to urban governance. Diverse social forces now participate in the urban governance process, moving from government-led management to collaborative, multi-actor governance (Huang & Zhang, 2016).

Based on the binding nature of decision-making, urban governance actors can be categorized as power-holding or non-power-holding entities. Power-holding actors include the municipal committees of the CPC and state organs at the municipal level (e.g., municipal people's congresses and their standing committees, municipal governments, municipal procuratorates, and courts). Non-power-holding actors include public institutions, enterprises, social organizations, and citizens. These diverse actors contribute to urban policy-making, help maintain social order, provide varied social services, and express and aggregate their interests, which are then conveyed to public authority organizations. Additionally, under China's unique context, urban governance actors can be classified into political, economic, and social organizations: (1) Political organizations include parties, state organs, and political groups established to achieve political leadership, public management, or public participation. (2) Economic organizations are profit-driven entities that utilize various production factors to offer goods or services in the market. (3) Social organizations are non-governmental and non-profit social groups formed to achieve public or mutual benefits (Yang, 2017).

1.3 Digital Urban Governance

With societal advancements, urban challenges have grown increasingly complex and diverse, significantly heightening the difficulty of urban governance. Fortunately, societal progress itself has brought new opportunities for urban governance, largely stemming from technological innovations, particularly the rapid development of new-generation information and communication technologies (ICT) over the past decade. From a literal perspective, **digital urban governance** can be understood in two ways: governance targeting digital or smart cities, or governance enhanced by digital means. This book adopts the latter perspective, defining digital urban governance as a new governance model that leverages technologies like big data, cloud computing, the Internet of Things (IoT), blockchain, and artificial intelligence (AI) to enhance governance capacity. Due to differing perspectives and preferences, digital urban governance is sometimes referred to as urban digital governance or smart urban governance in other studies. Some researchers consider digital urban governance an early stage of smart urban governance, with the latter encompassing richer connotations (Jiao & Shi, 2020). In this book, digital urban governance is viewed as a continuously evolving concept enriched through ongoing practices.

Sun and Shi (2020) identify three core components of smart urban governance: smart tools, smart models, and smart outcomes. Smart tools refer to the various hardware and software associated with new-generation information and communication technologies (ICT), including the Internet of Things (IoT), cloud computing, big data, and artificial intelligence (AI). These tools are fundamental to digital urban governance, characterized by their technical and instrumental features. Without the widespread application and support of digital tools, the concept of digital urban governance cannot be realized. Smart models represent new forms of human collaboration manifested in smart urban governance. Smart governance primarily achieves its objectives through external collaboration and participation, as well as through the intelligent transformation of internal organizational coordination and decision-making processes (Meijer & Bolívar, 2016). In traditional urban governance, one of the most challenging aspects is building collaborative relationships among governments, the public, and various stakeholders. By leveraging digital technologies, smart urban governance enables more effective arrangements and cooperation tailored to specific governance objects and goals. Several theories have been developed to explain smart urban governance, including collaborative governance, multi-actor governance, and participatory governance. Smart outcomes are the goals of making urban governance smarter and more digitalized. These outcomes are reflected in several areas, including: Enhancing the efficiency of public service delivery and resource utilization; Increasing transparency in governance processes and fostering public participation; Reducing environmental impact and improving urban ecological performance; Responding swiftly to urban disasters and developing balanced risk management strategies; Elevating residents' educational attainment and attracting highly educated human resources; Promoting online interactions among diverse stakeholders to enhance government service capabilities; and more.

Li (2020) notes three critical logics in the digital transformation of urban governance: data logic, technology logic, and governance logic. Data logic, with data as its carrier, originally referred to the extraction of information from objects. It later evolved to encompass the collection, storage, and analysis of big data, becoming one of the most defining characteristics of digital existence. Technology logic, with technology as its carrier, originally referred to tools for achieving human goals. It later evolved into technology as applied science. In the current digital context, the term "technology" primarily refers to the new generation of information and communication technologies. Governance logic, with public affairs governance as its carrier, initially referred to the joint actions of the government and various stakeholders to solve problems. It later evolved into a multi-level, multi-domain, and cross-departmental form of collaborative cooperation and networked action. Technology and data must be embedded within governance scenarios in order to be effective. Digital urban governance requires the effective integration of governance, technology, and data in order to improve and enhance the governance process.

2 The Evolution of Digital Urban Governance

2.1 E-Government

The origins of China's digital urban governance trace back to the 1980s with the advent of e-government. During the Sixth Five-Year Plan, the use of electronic computers in government administration was explicitly proposed. Some central departments, such as the Ministry of Finance, began establishing data centers to support data processing tasks. In 1987, the State Council convened the "National Government Office Automation Conference and Software Exchange Meeting" in Beidaihe, marking the beginning of government office automation in China.

In September 1993, the U.S. government announced the implementation of a high-tech plan known as the National Information Infrastructure. This plan aimed to establish an "information superhighway" modeled on the nascent Internet, enabling people to easily share vast information resources. To align with the global trend of constructing an information superhighway, China officially launched the "Three Gold Projects" at the end of 1993, consisting of the Golden Bridge, Golden Customs, and Golden Card projects: Golden Bridge focused on building a modern information infrastructure. Golden Customs refers to the electronic customs clearance business. Golden Card pertained to various card-based application systems, primarily focusing on electronic monetary applications. The "Three Gold Projects," led by the central government, represented a systemic effort to digitize government functions and marked the inception of China's government informatization (Shi et al., 2019). Its launch also signified the comprehensive rollout of the "Gold Projects." In August 2002, the "Guidelines on Advancing China's E-Government Construction" clarified the development of 12 key systems (i.e., the "Twelve Gold Projects"). These included refining four existing projects—office resource systems, Golden Customs, Golden Tax, and financial supervision (including Golden Card)—and initiating or accelerating eight others: macroeconomic management, Golden Finance, Golden Shield, Golden Audit, social security, Golden Agriculture, Golden Quality, and Golden Water systems.

The Qingdao Government Public Information Website, established in April 1998, was, from all perspectives, the first true government website in China. In January 1999, Beijing hosted the "Online Government Project Launch Conference," marking the official initiation of the Online Government Project. This initiative allowed the government to provide more convenient services, enabling citizens to complete administrative procedures remotely (Shi et al., 2019).

During this historical period, key projects such as office automation, the "Three Gold Projects," and the Online Government Project laid a solid foundation for the rapid development of China's e-government. After 2000, China introduced a series of policies to further promote e-government development, which led to new characteristics, such as: (1) Actively advancing information sharing and cross-departmental collaboration in government services; (2) Adapting to mobile internet development, with most government departments launching official microblogs, apps, WeChat public accounts, and Douyin accounts; (3) Responding to the national "Internet+" strategy and the need for administrative reform by streamlining service processes, innovating service models, promoting data sharing, breaking down information silos, and establishing integrated online government service platforms. Overall, the development of e-government has remained focused on governmental functions. While it is related to urban management, its focus differs.

2.2 Digital City

In January 1998, the U.S. Vice President Al Gore delivered a speech titled "The Digital Earth: Understanding Our Planet in the 21st Century" at the California Science Center. He introduced the concept of the Digital Earth, envisioning it as a virtual earth that integrates vast geographic data and presents it in three-dimensional, multi-resolution formats. By combining technologies such as Geographic Information Systems (GIS) and Virtual Reality (VR), the Digital Earth could store global geographic information on computers and create virtual geospatial environments.

In 2005, Google launched Google Earth, one of the most representative Digital Earth applications. It overlays geographic data and aerial imagery onto 3D models to showcase the entire planet. Users can explore the Earth using keyboards and mice, mark locations, and add text, photos, and videos for personalization. By 2019, Google Earth had captured satellite images of 36 million square miles (approximately 93.24 million square kilometers), covering areas inhabited by 98% of the global population. It also offered street view services, featuring 10 million miles (approximately 16.09 million kilometers) of 360° panoramic images (Nieva, 2019).

The concept of Digital Cities is considered a subset of Digital Earth, focusing on the digitization of geographic spaces within single administrative city regions. Its core technologies include spatial information technologies such as remote sensing, Geographic Information Systems, and satellite positioning systems. These technologies have been widely applied in urban planning, construction, and management. For instance, GIS can create virtual 3D city models, allowing management departments to visualize and manage the real world intuitively.

The concept of Digital Earth also attracted significant attention from Chinese scholars. In November 1998, the Earth Science Division of the Chinese Academy of Sciences held a symposium in Beijing, attended by numerous academicians and experts, to discuss Digital Earth. In May 2006, the International Society for Digital Earth was established in Beijing, making it one of the few international academic organizations headquartered in China (Jin & Guo, 2016). Many GIS companies in China have also developed products and solutions to support Digital City construction. For example, Beijing SuperMap Software Co., Ltd., China's largest GIS software provider, specializes in the development and application of GIS software technologies. Its SuperMap GIS software ranked first in market share within China's GIS software application market and has been widely applied in industries such as natural resources, municipal services, ecology, agriculture, transportation, and energy. Many cities have also undertaken Digital City construction projects. In 2006, China launched the "Demonstration Project for Building Digital Regional Geospatial Frameworks." Over the following five years, approximately 120 cities and districts initiated Digital City geospatial framework construction projects, establishing information systems for traffic management, municipal services, public security, underground pipelines, population management, land management, and urban renewal (Li & Xu, 2017).

2.3 Smart City

In November 2008, Sam Palmisano, CEO of IBM (International Business Machines), delivered a speech titled "*Smarter Planet: The Next Leadership Agenda*" at the Council on Foreign Relations in New York City, introducing the concept of a "Smarter Planet." In January 2009, IBM presented a report on the Smarter Planet at the U.S. Business Roundtable, aiming to stimulate the economy and establish America's future competitive edge by increasing investments in next-generation information infrastructure, such as broadband networks. U.S. President Obama responded positively to IBM's Smarter Planet concept, incorporating it into the U.S. national strategy and positioning it as a new economic growth point to address the financial crisis. IBM also took a series of steps to enter China's massive smart city market. In February 2009, IBM proposed the "Smarter Planet Wins in China" strategy and recommended focusing on the construction of smart power, healthcare, cities, transportation, supply chains (logistics), and banking industries. In August 2009, IBM released the *Smarter Cities in China* report.

The introduction of the "Smarter Planet" concept is considered a well-planned and strategically deployed initiative by IBM, given its global leadership in the key technologies (sensor networks, cloud computing, supercomputing, data mining, etc.) essential for realizing the Smarter Planet vision, which gives IBM the opportunity to regain a competitive advantage in the future (Wang & Yan, 2010). Despite this, the Smarter Planet concept quickly gained popularity worldwide and significantly influenced the path of future urban development.

Similar to the relationship between the concepts of digital earth and digital cities, smart cities are a concept relative to Smarter Planet. Specifically, a smart city refers to the use of next-generation information and communication technologies such as the Internet of Things (IoT), cloud computing, and big data to enhance the intelligence of urban planning, construction, management, and operations, thereby improving the quality of urban living environments and promoting sustainable urban development. Following the emergence of the smart city concept, China also experienced a boom in smart city construction.

According to the *China Smart City Standardization White Paper*, which outlines the technical framework for smart cities, the smart city includes the IoT perception layer, the network communication layer, the data and service support layer, and the smart application layer (China Electronics Standardization Institute, 2013) (see Fig. 1).

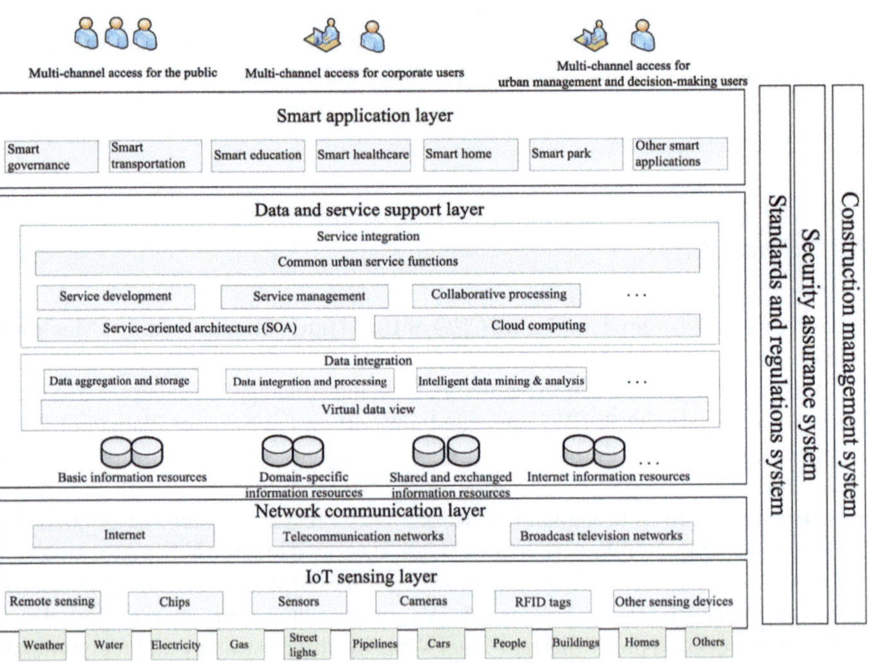

Fig. 1 A smart city technology model. *Source* China Electronic Standardization Institute (2013)

(1) IoT Perception Layer: This layer primarily provides the intelligent perception capabilities for the urban environment. Centered on IoT technology, it utilizes chips, sensors, radio frequency identification (RFID), cameras, and other technical means to collect, monitor, and control various types of information from urban infrastructure, environment, and buildings.
(2) Network Communication Layer: This layer includes ubiquitous, shared, convenient, and high-speed network communication infrastructure. It focuses on the construction and integration of the Internet, telecommunications networks, and broadcasting networks to create a city-wide, high-capacity, high-bandwidth, and highly reliable optical network and wireless broadband network.
(3) Data and Service Support Layer: This is the core component of smart city construction. It enables the aggregation, sharing, and utilization of urban information resources and supports various smart applications. This layer primarily involves the application of technologies such as cloud computing, big data, and service-oriented architecture (SOA).
(4) Smart Application Layer: This layer refers to the various smart applications built on top of the previous three layers. By developing a wide range of smart applications aimed at governments, enterprises, and residents, it provides more refined and intelligent services to different stakeholders. The construction of the application layer can promote the development of various industries, such as smart governance, transportation, education, healthcare, home automation, and industrial parks. It can provide comprehensive information services to the public, enterprises, and city managers, promote intelligent operation, efficient management, and universal services, and drive the establishment and development of modern industrial systems in cities.

Ningbo is one of the earliest cities to propose the development of a smart city. In September 2010, the Ningbo Municipal Party Committee and the People's Government of Ningbo issued the "*Decision on the Construction of Smart Cities*," outlining a smart city application system that includes ten major areas: smart logistics, manufacturing, trade, energy, public services, social management, transportation, healthcare, housing, and culture. The plan also called for the development of six key smart industry bases: a network data base, a software research and development and promotion industry base, a smart equipment and product research and manufacturing base, a smart service industry demonstration and promotion base, a smart agriculture demonstration and promotion base, and a smart enterprise headquarters base.

In March 2012, the Beijing Municipal Government issued the "*Action Outline for Smart Beijing*," which stated that Beijing would "build ubiquitous, integrated, intelligent, and trustworthy information infrastructure, realize precise population management, intelligent traffic supervision, scientific resource allocation, and secure urban management systems, and establish a digital living environment that is integrated, personalized, and humanized, covering both urban and rural residents and accompanying citizens throughout their lives. The goal was to popularize a new type of enterprise operation model that deeply integrates information technology with industrialization and leads corporate innovation and transformation. It also aimed to create a government service system that is efficient and citizen-centered, facilitating the deep integration of information technology with all aspects of urban economic and social development, ultimately achieving world-class levels of informatization and transitioning from 'Digital Beijing' to 'Smart Beijing'."

To standardize and promote the healthy development of smart cities, the Ministry of Housing and Urban-Rural Development officially released a "*Notice on Carrying Out National Smart City Pilot Work*" in December 2012. Along with this, two documents were issued: the "Interim Management Measures for National Smart City Pilots" and the "National Smart City/District/Town Pilot Indicator System (Trial)." Between 2013 and 2015, the Ministry of Housing and Urban-Rural Development, along with the Ministry of Science and Technology, announced 290 national smart city pilot projects in three batches.

In August 2014, the National Development and Reform Commission and eight other ministries jointly issued the "Guiding Opinions on Promoting the Healthy Development of Smart Cities," which systematically elaborated on the guiding principles, basic objectives, and institutional designs for smart city construction. The document emphasized that "by 2020, a batch of distinctive smart cities should be established." With the support of various policies, the construction of smart cities in China has smoothly started.

In April 2016, during a work conference on cybersecurity and informatization, General Secretary Xi Jinping stated that "informatization should be used to advance the modernization of the national governance system and governance capabilities, coordinate the development of e-government, create an integrated online service platform, promote the construction of new smart cities in a graded and classified manner, break down information barriers, and build a national information resource-sharing system. This will help us better use informatization to perceive social dynamics, smooth communication channels, and assist in scientific decision-making."[1] From that point onward, China entered the construction phase of new-type smart cities.

In December 2016, the State Council officially released the *13th Five-Year Plan for National Informatization*, which outlined the action goals for the construction of new-type smart cities: "By 2018, 100 new-type exemplary smart cities will be built in a graded and classified manner; by 2020, significant progress will have been made in smart city construction, forming ubiquitous public services, a transparent and efficient online government, an innovative information economy, precise and fine-tuned urban governance, and a secure and reliable operational system." The *14th Five-Year Plan for National Informatization*, released in December 2021, further stated that, in promoting the high-quality development of new-type smart cities, it is necessary to "advance the integrated development of smart city clusters based on local conditions, focusing on key areas such as public transportation, express logistics, medical care and education, urban management, environmental protection, license management, market supervision, public safety, and emergency management. This will promote a series of smart application regions to collaborate and coordinate,

[1] Xi Jinping. Speech at the National Cybersecurity and Informatization Work Conference. *People's Daily*, April 26, 2016 (2).

fostering the coordinated development of regional informatization." The new-type smart cities aim to guide regions in utilizing the Internet, big data, and artificial intelligence to advance urban governance and public service intelligence.

In April 2020, the *Shandong Province New-Type Smart City Construction Indicators*, organized by the Shandong Provincial Big Data Bureau and compiled by the Shandong Academy of Sciences' New Generation Technology Standardization Research Institute, was officially released. This set became the first provincial-level local standard for the graded and classified advancement of new-type smart city construction in China. The standard includes city-level, county-level, and community-level indicators. The city and county standards cover six primary indicators: digital welfare, digital government, digital economy, infrastructure, guarantee measures, and local characteristics. Based on differences in economic levels, population sizes, and tasks, the standards define 49 secondary indicators for cities and 44 for counties. Each secondary indicator is divided into five levels, ranging from 1 to 5 stars. The smart community indicators cover four primary indicators: infrastructure, community services, community safety, and service innovation, along with 24 secondary indicators.

2.4 Digital Urban Governance

Although focused on "urban management," grid-based management can be seen as a representative practice of early digital urban governance, hailed as "a breakthrough in the integration and innovation of modern urban management" (Chen, 2006). In 2004, Dongcheng District of Beijing was the first in the country to implement the grid-based urban management model. This model divides each street into multiple grids, each covering an area of 10,000 square meters, with a dedicated grid worker responsible for each grid. Based on this, an urban informatization management system encodes the urban components within each grid, which include various elements related to urban management, such as manhole covers, streetlights, mailboxes, and telephone booths. Grid-based management also restructures the management system by separating supervision from command, ensuring clear division of responsibilities, and facilitating evaluation and assessment. In December 2015, the Central Committee of the Communist Party of China and the State Council issued the "*Guiding Opinions on Deepening Urban Law Enforcement System Reform and Improving Urban Management Work*," which required that by the end of 2017, all cities and counties nationwide must integrate and form digital urban management platforms.

In the context of advancing the modernization of national governance systems and governance capabilities, digital urban governance has developed rapidly in the second decade of the twenty-first century. In November 2013, the Third Plenary Session of the 18th Central Committee of the Communist Party of China proposed, "the overall goal of comprehensively deepening reform is to improve and develop the socialist system with Chinese characteristics and advance the modernization of the national governance system and governance capabilities." In October 2019, the Fourth Plenary

Session of the 19th Central Committee of the Communist Party emphasized that the national governance system and governance capabilities are the concentrated embodiment of the socialist system with Chinese characteristics and its execution ability. Adapting to the changes in the primary contradictions of society and overcoming the risks and challenges on the path forward requires efforts in modernizing the national governance system and governance capabilities. The construction of the national governance system and governance capabilities has significant theoretical and practical implications for all aspects of China's social development. During his visit to Shanghai in November 2018, General Secretary Xi Jinping emphasized that "urban governance is an important part of the modernization of the national governance system and governance capabilities. First-class cities require first-class governance and should focus on scientific, refined, and intelligent management."[2]

After years of exploration, there are now three typical cases of digital urban governance: Shanghai's "One Network for All Services, One Network for All Management," Hangzhou's City Brain, and Xiong'an's Digital Twin City. Compared to the earlier stage of grid-based management, the current digital urban governance practices emphasize "governance" and make full use of the new generation of information and communication technologies.

Shanghai's "One Network for All Services, One Network for All Management": During the 2019 China International Import Expo in Shanghai, General Secretary Xi Jinping visited the city and put forward the important idea that "people's cities built by the people and the people's cities for the people." He emphasized the need for both "One Network for All Services" and "One Network for All Management" in urban governance, regarding the construction of this one network of two aspects as key projects to improve the city's modern governance capabilities and levels (Zhao, 2020). "One Network for All Services" connects the information systems of different government departments, allowing citizens to complete multiple tasks that require the involvement of various government departments by logging into the government's integrated portal, using the Suishenban app, or accessing self-service kiosks. This eliminates the need for citizens to travel between different administrative departments. "One Network for All Management" refers to the ability to present holographic and panoramic views of all elements, objects, processes, and results of urban governance on a single platform, enabling integrated and closed-loop management of urban governance matters. The construction of these two aspects of One Network has helped explore a new path for the governance of super-large cities with Chinese characteristics. Since its launch, "One Network for All Services" has attracted 21.86 million registered users, covering 2,312 service items. This has reduced processing time by 59.8%, decreased paperwork by 52.9%, and achieved a satisfaction rate of 99.7%, greatly enhancing the sense of benefit among citizens and enterprises (Wu & Li, 2020).

[2] Strengthening Confidence and Determination for a New Leap in Reform and Opening Up: Accelerating the Enhancement of Urban Capabilities and Core Competitiveness. *People's Daily*, November 8, 2018 (1).

Hangzhou City Brain: Hangzhou has been at the forefront of digital urban governance in China. The City Brain project began in April 2016, initially focusing on traffic management as a breakthrough area to use big data to improve urban traffic. Today, it has evolved from traffic congestion management to overall urban governance, with many phased achievements. The application scenarios of Hangzhou's City Brain are continuously expanding, and now include 48 digital application scenarios, such as health codes, talent codes, online sincere-upright services for enterprises, digital urban management, digital cockpit, and "Two Mountains Bank," with 113 application projects released in three batches. Since 2016, the City Brain has transitioned from addressing traffic congestion to tackling city-wide governance, showing vast potential in digital-enabled urban governance, particularly during the COVID-19 pandemic, where it played a critical role. In March 2020, General Secretary Xi Jinping visited the City Brain's operation command center in Hangzhou to learn about the use of health codes, cloud services, and other methods to advance pandemic control and the resumption of work and production. He emphasized that the City Brain is an important initiative in building "Digital Hangzhou" and that cities can become "smarter" by leveraging big data, cloud computing, and artificial intelligence to modernize urban governance. From informatization to intelligence, and finally to wisdom, is the necessary path to building smart cities.[3] In June 2020, the CPC Hangzhou Municipal Committee decided to strengthen and optimize the City Brain, aiming to build it into a "key window" for the construction of new-type smart cities in the country. The goal is to use cutting-edge technologies such as big data, cloud computing, blockchain, and AI to innovate urban management methods, models, and concepts, further modernizing urban governance systems and capabilities. Hangzhou plans to leverage the City Brain as a key tool to enhance its urban governance modernization and create a comprehensive digital governance system, positioning itself as the "No. 1 city for digital governance in China."

Xiong'an Digital Twin City: The most important feature of the digital construction of Xiong'an is that while building a real city, it also gathers multi-source data through the Internet of Things to create an intelligent model for the city of the future, forming a digital twin city that grows in parallel with Xiong'an. This makes it the first fully digital city in the world, built from the very beginning. The digital twin city applies the concept of digital twins at the urban level, enabling the physical world to correspond, map, and interact with a virtual space. It constructs a complex system where virtual and physical cities are synchronized in planning and construction. The digital twin city aims to achieve full digitalization and virtualization of urban elements, real-time and visualized urban status, collaborative and intelligent urban governance, and synchronized planning and construction of both virtual and physical cities. The Xiong'an digital twin city will provide a model for building future cities that are green, intelligent, and livable, and will offer new ideas and solutions for digital urban governance.

[3] Zhang Xiaosong, Yang Weihan, and Zhu Jichai. Xi Jinping: Using Informatization to Make Cities "Smarter". *Xinhua Net*, April 1, 2020.

Overall, the development of e-government, digital cities, and smart cities has different emphases: e-government focuses on the application of digital technology in public affairs, while digital cities and smart cities focus on the use of digital technology across the entire process of urban planning, construction, and management. However, because they share a close connection with digital urban governance, the development of digital urban governance is closely related to e-government, digital cities, and smart cities. In a sense, digital urban governance can trace its origins to e-government, which began in the 1980s, and has evolved along with the development of digital cities and smart cities. In the current era of advancing the modernization of national governance systems and governance capabilities, digital urban governance has become an independent academic field and a practical hotspot.

3 Theoretical Foundations of Digital Urban Governance

A few studies have attempted to explore digital urban governance from a theoretical perspective. Razaghi and Fingerm (2018), based on the socio-technical systems theory, studied smart urban governance from both technological and managerial dimensions, proposing that technological innovation should be used to address the social and political challenges brought about by smart city development. Han and Li (2019) argue that smart governance is a governance form that aligns with urban space, comprising three elements: "actors," "technical components," and "rule systems." Smart governance achieves clarity of governance subjects, objects, processes, and outcomes through the data encoding, processing, and computing of social facts. Sun and Shi (2020), after clarifying the evolution of the concept of smart cities, state that smart cities can reshape government functions, and smart governance can drive innovations in public services and lead reforms in government decision-making. Wu et al. (2020), using the Hangzhou City Brain as a case study, explored the practical models of urban governance transformation in the digital age based on multi-level governance theory. We hold that the model's characteristics include the holistic nature of digital governance, a clear hierarchical structure, and a flexible, resilient organizational system. Although this study did not directly approach digital governance theory, we argue that digital governance has transformed the multi-level governance model, particularly in terms of the networked restructuring of decentralized governance structures and the dissolution of boundaries between multi-level governance systems. Li (2020) proposed a new framework to understand the digital transformation of urban governance. This framework, based on a dual-layer nested model aimed at both citizens and decision-makers, enables effective integration between multiple levels of government, different governmental departments, and the needs of citizens, ultimately achieving the governance goal of "dual-layer interfaces, multiple integrations, and integrated supply-demand." Based on a review of existing literature, it can be inferred that, as digital governance in the urban context, the theoretical background of digital urban governance is rooted in digital governance theory, which itself originates from reflections on new public management theory and the development of information technologies.

3.1 New Public Management Theory

Since the 1970s, in response to fiscal crises, efficiency crises, and other challenges in government administration, Western countries initiated a reform movement aimed at privatizing government management, prioritizing efficiency, and adopting a customer-oriented approach. Among these reforms, New Public Management (NPM) theory has been the most widely applied, as it advocates for the use of business-oriented, market-driven management methods to reform the public sector, effectively addressing the dilemmas faced at that time. New Public Management is based on economics and centers on coordinating the relationship between government and market. Its theoretical roots can be traced back to public choice theory, transaction cost theory, and agency theory in economics (Zhou, 2005).

New Public Management emphasizes building a customer-oriented government. The government should not be a bureaucratic entity distant from the people or above society. A good government should be entrepreneurial, with government officials acting as responsible managers and citizens being customers who pay taxes in exchange for government services. The government should serve citizens with a customer-oriented mindset, striving to improve the quality of public services.

New Public Management also advocates for the objective and accurate evaluation of government department performance through scientific methods. The evaluation mainly covers aspects such as cost-benefit analysis, customer satisfaction, and input-output ratios. The 3E criteria—economy, efficiency, and effectiveness—serve as the primary standards for performance evaluation in practice and are considered the key value goals for government policymaking. New Public Management theory advocates that the allocation of resources in government management should be linked to the performance of management personnel. It stresses that compensation should be based on performance rather than traditional tasks and that a budget system should be established that allocates funds based on outcomes rather than inputs (Yang, 2005).

New Public Management theory also supports the introduction of market mechanisms, encouraging competition between the public and private sectors as well as within the public sector, to reduce the size of government and improve the efficiency of public service provision. Although the government is naturally the provider of public services, NPM argues that not all public services should be provided by the government. It advocates for private sector participation in public service delivery, establishing partnerships between the government and private sector. By introducing market mechanisms, it helps alleviate fiscal pressure on the government, compensates for the shortcomings of government as the sole provider of public goods, and thus improves the operational mechanisms and efficiency of government.

3.2 Digital Governance Theory

However, as previously mentioned, New Public Management (NPM) theory is grounded in economics and emphasizes applying business management approaches and methods to the public sector, encouraging competition and incentives, while lacking a solid theoretical foundation in public administration. This poses significant risks because the public and private sectors are fundamentally different: the public sector emphasizes public values such as democracy and fairness, while the private sector seeks to maximize profits. Introducing market mechanisms into the public sector may be beneficial in the short term, but it could have significant negative consequences in the long term (Ma, 2016). Zhou (2005) identifies five major issues with New Public Management theory: (1) Replacing public sector management with business management leads to a lack of publicness, thereby undermining the legitimacy of public administration; (2) It weakens values such as fairness and democracy; (3) Transferring responsibilities previously held by the government to society may result in the loss and reduction of the government's public responsibilities; (4) Comparing citizens to customers is inappropriate because the relationship between the government and citizens differs from that between producers and consumers; (5) Outsourcing government services to private enterprises may lead to corruption issues. Dunleavy (2005) argues that the dominance of New Public Management theory has greatly increased the complexity of institutions and policies, and its decline is inevitable.

While New Public Management theory has gradually waned, information technology has experienced rapid growth and has had a profound impact on public administration. Weng (2019) argues that, from the perspective of government functions, information technology has facilitated a shift from a "government-centered" to a "citizen-centered" approach and from a "regulatory society" to a "service society." Government departments began to use information technology to upgrade services, especially in reshaping the four key relationships: government-to-government (G2G), government-to-employee (G2E), government-to-business (G2B), and government-to-citizen (G2C). From the perspective of government organizational structure, information technology has enabled various levels of government to have unified network service platforms, facilitating information sharing, reducing intermediate organizational layers, enhancing internal communication and coordination, and raising the information literacy requirements for government staff. From the perspective of government operations, the development of e-government has greatly enhanced government operational efficiency. Initially, e-government focused on office automation and management information systems within government departments, gradually expanding to use network technologies to promote coordination and information sharing between departments, provide unified government services to society, and establish one-stop government service platforms.

Digital governance theory emerged against the backdrop of reflections on New Public Management and the rapid development of information technology. This theory originated in the 1990s, with Patrick Dunleavy from the London School of

Economics being one of its leading scholars. Xu and Liu (2006) suggest that digital governance can be understood from both narrow and broad perspectives. In a narrow sense, digital governance refers to a governance model that uses information and communication technologies to simplify government administration and enhance democracy in interactions between the government and citizens, businesses, and within the government itself. In a broader sense, digital governance is not just the simple application of information technology to public affairs; it is a form of social-political organization and activity related to the organization and utilization of political and social power, which includes comprehensive governance of economic and social resources and involves various activities that impact government, legislatures, and the public administration process.

Digital governance theory encompasses three major themes: reintegration, needs-based holism, and digitization changes. Below is an introduction to each theme and its constituent elements.

1. **Reintegration**

Reintegration is proposed in response to the excessive fragmentation issues found in New Public Management (NPM) theory. It emphasizes reclaiming and reintegrating functions that were separated under NPM. Through scientific planning, reintegration seeks to reduce resource duplication and waste, ease the burden on citizens, and enhance the accessibility of public services. Specifically, reintegration consists of the following nine elements:

(1) Rollback of Agencification and Fragmentation. This element advocates for integrating government agencies with similar functions and establishing collaborative structures at the community level.
(2) Joined-up Governance. Through the restructuring and merging of key departments within the central government, joined-up governance aims to address the fragmentation and inefficiencies brought about by the NPM movement. Joined-up governance can effectively counter the power expansion issues under traditional one-way governance, while also leveraging citizen oversight.
(3) Re-governmentalization. Re-governmentalization refers to the process of returning functions that were previously outsourced to the private sector back to the public sector for management. It is not merely a return of functions but involves the government taking a comprehensive view of public affairs, utilizing information technology to formulate public policies and provide quality public services and products.
(4) Reinstating or Re-strengthening Central Government Processes. Due to the excessive and dispersed competition caused by NPM, some countries have begun to re-strengthen central government processes. For instance, the UK government has strongly supported central electronic reform projects, starting in 2005, to reduce redundant information and avoid an overabundance of government websites in digital governance.

(5) Radically Squeezing Production Costs. This mainly involves cutting the number of civil servants, especially in government departments heavily reliant on information technology. The primary goal is to achieve rational resource utilization and increase the allocation of resources to frontline public service workers.
(6) Re-engineering Back-office Functions. On one hand, new information technologies are used to enhance productivity; on the other hand, back-office functions are redesigned by streamlining processes, such as creating demand-driven production lines.
(7) Procurement Concentration and Specialization. One issue caused by NPM is the decentralization and lack of specialization in procurement. For example, in the NPM era, procurement functions in the UK were dispersed across multiple departments and agencies. By consolidating these functions from 270 departments and agencies into a few major procurement centers, procurement could be both centralized and specialized, potentially saving up to £20 billion (Peter, 2004).
(8) Shared Services on a Mixed-Economy Basis. This involves encouraging smaller departments and agencies to share services in order to reduce organizational redundancies.
(9) Network Simplification. By integrating functions back into central departments, this element seeks to reduce additional administrative costs across multiple agencies, fully leveraging online resources from public sector institutions.

2. **Needs-Based Holism**

Needs-based holism seeks to simplify and adjust the relationship between government agencies and citizens, reduce administrative costs, and minimize unnecessary procedures, creating a more comprehensive and streamlined government management process. Holism consists of seven key elements:

(1) Interactive Information-Giving and Seeking. This is the most fundamental element, promoting interactive mechanisms such as call centers and online electronic services to encourage public servants to respond to citizens' needs and preferences from a holistic perspective.
(2) Client-Based or Needs-Based Reorganization. This involves reorganizing institutions around the needs of a specific group of clients. For example, the UK government established a Pension Service under the Department for Work and Pensions to handle all welfare matters related to the elderly.
(3) One-Stop Provision. A new service model widely adopted by local government portals and regional administrative service centers. One-stop services include administrative services provided by one-stop shops, one-stop windows, and integrated services via the internet. This model encourages governments to reuse information previously collected, avoiding redundant data collection, and preventing resource waste.
(4) Data Warehousing. Data warehousing provides effective computing resources through computer hardware and software, the internet, and government intranets. It enables precise data analysis, offering data and technical support

for government decision-making. Collecting large amounts of welfare, tax, and other information into databases can significantly enhance the convenience of accessing and utilizing such information.

(5) End-to-End Service Re-engineering. This measure encourages government departments to handle matters from a holistic perspective. For instance, Canada has used information technology to significantly reduce the number of application forms needed during the subsidy application process.
(6) Agile Government Processes. This aims to enhance the responsiveness and flexibility of government decision-making processes.
(7) Sustainability. Ensures that the development of the public sector does not have negative environmental impacts. This aligns with the needs-based holistic approach, as environmental protection and sustainability are among the key concerns of citizens and environmental organizations.

3. **Digitization Changes**

Digitization changes are not merely a supplement to traditional public administration but represent a complete and transformative shift. The impact is primarily reflected in changes to government organizational structures and internal culture, while also promoting shifts in citizens' attitudes towards government use of information technology. Digitization changes consist of the following eight elements:

(1) Electronic Service Delivery and E-Government. Electronic service delivery enables the government to convert most paper-based administrative processes into electronic workflows. With globalization and informatization, e-government has made significant progress, offering advantages such as ease of operation, strong practicality, and efficient management.
(2) Web-Based Utility Computing. Smaller government agencies can purchase computing or information technology services on demand from a marketplace with multiple suppliers or select services from menus provided by primary suppliers.
(3) New Forms of Automated Processes. Emphasizes the use of zero-touch technology, where the majority of tasks in administrative affairs are handled by computers without human intervention.
(4) Radical Disintermediation. This refers to a new trend in communication between the government and businesses or citizens, where the government uses internet platforms to communicate directly with the public and enterprises, bypassing the traditional "gatekeeper" role of civil servants and institutional staff.
(5) Active Channel Streaming and Customer Segmentation. Utilizes digital services to shift digital governance targets into the realm of digital governance, creating highly differentiated, question-driven electronic systems for customer segmentation.
(6) Mandated Channel Reductions. Requires businesses or individuals to transact with government agencies through electronic payments or other digital channels, particularly in key service areas such as taxation.

(7) Facilitating Isocratic Administration. Emphasizes a shift from government-centric to citizen- and enterprise-centric interactions during exchanges between citizens, businesses, and government agencies.
(8) Moving Towards Open-Book Government. In the digital age, governments are transitioning from a closed state to an open one, making it easier for citizens to self-manage and creating interconnected systems between the government, enterprises, and citizens.

Dunleavy and Margetts (2010) divide the development of digital governance theory into two phases: from 2002 to 2010, and post-2010. In the second phase, with the development of new information communication technologies like big data and cloud computing, digital governance theory has undergone significant changes. For example, from the reintegration perspective, the major trend is the increasing adoption of intelligent centers and decentralized delivery systems by governments, a concept largely borrowed from private enterprises like Walmart. From a holistic perspective, the social security system is undergoing a new round of integration, shifting towards online forms, while the emergence of citizen single accounts has effectively enhanced online access rates and government-citizen interaction efficiency. From the digitization perspective, governments are starting to build super websites and reduce the overall number of government websites, and are also building government clouds for free data storage, comprehensive data retention, and developing various government apps to foster interconnected interactions (Wang, 2019).

It should be noted that although digital governance theory has gradually expanded into urban governance, the value demands, practical paths, and digital governance aspects of smart cities share similarities. However, few scholars have systematically explored smart cities from the perspective of digital governance theory (Han & Ma, 2016). How to systematically interpret urban governance from the standpoint of digital governance theory remains an important issue in the current rapidly evolving era of information technology.

4 Organization of the Book

This book attempts to understand digital urban governance from two perspectives: technology-enabled governance and data-driven governance (see Fig. 2). On the one hand, new-generation information and communication technologies, such as the Internet of Things (IoT), cloud computing, big data, and artificial intelligence, are being rapidly applied to urban governance. These technologies integrate into every phase of urban planning, construction, management, and operations, serving multiple stakeholders, including governments, enterprises, social organizations, and citizens. The use of information technology allows for comprehensive data collection from cities, enabling a more thorough understanding of urban conditions, enhancing communication channels between governments, businesses, and citizens, and fostering innovation in the operational mechanisms of urban governance. These advancements contribute to the improvement of urban governance capabilities. On the other hand, data is often viewed as a treasure trove akin to "new oil." Cities are continuously generating data, and the widespread adoption of IoT devices has greatly expanded the ways in which urban data can be collected. The construction of big data centers has made it possible to receive and store massive amounts of real-time city data. This wealth of data helps us scientifically understand the behavior of governance entities, such as governments, businesses, social organizations, and citizens, as well as the state characteristics of governance objects, such as buildings, roads, and green spaces. Such data creates the conditions for improving urban governance. Moreover, the vast amount of real-time data enables the prediction of human behavior, facilitating targeted policy measures and providing precise, efficient, and convenient services. In contrast to traditional urban governance processes, data-driven urban governance practices place greater emphasis on enhancing the collection, processing, analysis, and decision-support capabilities of data.

Chapter 2 introduces new-generation information and communication technologies and their empowering potential for urban governance, specifically focusing on IoT, big data, cloud computing, mobile internet, artificial intelligence, and blockchain

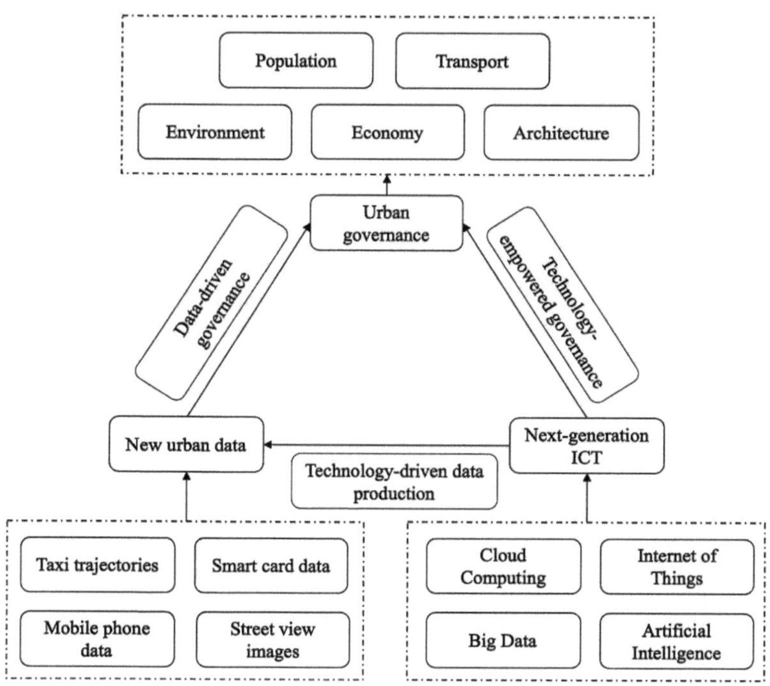

Fig. 2 The organizational framework of digital urban governance used in this book

technology. Chapter 3 discusses the new types of urban data and their potential applications in urban governance, mainly covering smart card transaction data, mobile phone data, point-of-interest data, social media data, and trajectory data.

Chapter 4 uses the examples of the City Brain and health codes to illustrate how technology empowers urban governance. Correspondingly, Chap. 5 uses traffic big data (such as metro card swipe data and user cycling big data) and mobile phone signaling data to show how data-driven urban governance research can be conducted.

In the final chapter, we summarize the entire book, discuss the potential challenges that may arise in the process of digital urban governance, and look forward to the future of digital urban governance.

References

Chen, P. (2006). *A new model for grid-based urban management*. Peking University Press.

China Electronics Standardization Institute. (2013). *White paper on the standardization of smart cities in China*. Retrieved April 10, 2021, from http://www.cac.gov.cn/2014-09/28/c_1112661626.htm.

Dunleavy, P. (2005, April 6). New public management is dead—Long live digital-era governance. *Journal of Public Administration Research and Theory, 16*(3), 467–494. https://academic.oup.com/jpart/article-lookup/doi/10.1093/jopart/mui057.10.1093/jopart/mui057

Dunleavy, P., & Margetts, H. (2010, April 6). The second wave of digital era governance. *APSA 2010 Annual Meeting Paper*. https://papers.ssrn.com/abstract=1643850

Gao, Q. Q., & Que, T. N. (2020). The space and prospects of blockchain in urban governance. *E-Government, 1*, 84–91.

Han, Z. M., & Li, C. S. (2019). Clarity and technical logic of urban governance: An analysis centered on smart governance. *Exploration, 6*, 44–54.

Han, Z. Z., & Ma, W. J. (2016). Exploration of digital governance theory and its application. *Public Administration Review, 1*, 92–109.

Huang, Q. H., & Zhang, Q. Z. (2016). *Research on the modernization of the urban governance system and capacity in Xiamen*. Social Sciences Academic Press.

Jiao, Y. L., & Shi, C. (2020). From digital urban management to intelligent urban governance: A study of the paradigm shift in urban governance in China. *Fujian Forum (Humanities and Social Sciences Edition), 11*, 37–48.

Jin, J. J., & Guo, Y. L. (2016). *Smart cities: Urban governance in the age of big data and the internet*. Electronics Industry Press.

Li, C. H., & Xu, H. Z. (2017). *Introduction to smart cities*. Social Sciences Academic Press.

Li, W. Z. (2020). Constructing the dual-layer nested governance interface: The direction and path of the digital transformation of urban governance. *E-Government, 7*, 32–42.

Ma, W. J. (2016). *Research on digital governance theory and its applications* (Unpublished master's thesis). Yanshan University, Qinhuangdao.

Meijer, A., & Bolívar, M. P. R. (2016, April 10). Governing the smart city: A review of the literature on smart urban governance. *International Review of Administrative Sciences, 82*(2), 392–408. http://journals.sagepub.com/doi/10.1177/0020852314564308.10.1177/0020852314564308

Nieva, R. (2019, March 1). *Google Maps has now photographed 10 million miles in Street View*. https://www.cnet.com/news/google-maps-has-now-photographed-10-million-miles-in-street-view/

Northam, R. M. (1979). *Urban geography*. Wiley.

Peter, G. (2004, April 10). *Releasing resources to the front line: Independent review of public sector efficiency*. HM Treasury. https://scholar.google.com/scholar?hl=zh-CN&as_sdt=0%2C5&q=Releasing+Resources+to+the+Front-Line%3A+Independent+Review+of+Public+Sector+Efficiency.&btnG=

Razaghi, M., & Fingerm, A. (2018). Smart governance for smart cities. *Proceedings of the IEEE, 106*(4), 680–689.

Shi, L. J., Zhang, Z. H., & Zhang, J. W. (2019). *E-Government*. People's Posts and Telecommunications Publishing House.

Sun, T. Y., & Shi, W. B. (2020). Research on international academic frontiers in smart urban governance. *Housing Technology, 11*, 1–8.

Wang, S. Q. (2019). The theory of governance in the digital age: Background, content, and brief evaluation. *Foreign Social Sciences, 2*, 96–104.

Wang, Y. H., & Yan, B. (2010, June 17). The dangerous Internet of Things. *Finance and National Weekly*. Retrieved March 1, 2021, from https://www.reuters.com/article/idCNCHINA-2486220100617

Weng, S. H. (2019). Theory of governance in the digital age: New responses from Western government governance and their implications. *Comparative Economic and Social Systems, 4*, 138–147.

Wu, K., & Li, Z. G. (2020, May 17). Shanghai: One network for all services, one network for comprehensive management. *Economic Daily*. Retrieved April 11, 2021, from http://news.jschina.com.cn/scroll/guonei/202005/t20200517_2552858.shtml

Wu, W. Q., Zhou, J. X., & Xie, N. N. (2020). The transformation of urban governance: Multi-level governance in the digital age. *Journal of Zhejiang University (Social Science Edition), 1*, 54–60.

Wu, Z. Q., & Li, D. H. (2010). *Principles of urban planning*. China Architecture and Building Press.

Xu, X. L., & Liu, Y. (2006). The impact of digital governance on good governance in urban governments. *Public Administration Journal, 3*(1), 13–20.

Yang, H. S. (2017). *Urban governance in transition*. Renmin University Press.

Yang, M. W. (2005). A review of new public management theory. *Journal of Sichuan Administrative Institute, 2*, 21–24.

Yang, X. Y., & Zhang, T. T. (2020). Knowledge map, research hotspots, and prospects for urban governance research in China: A perspective based on keyword co-occurrence clustering. *Modern Urban Research, 7*, 123–130.

Ye, C., Yu, J., Zhang, Q. Y., et al. (2021). From governance to urban-rural governance: International frontiers, development trends, and China's path. *Progress in Geography, 40*(1), 15–27.

Zhao, Y. (2020). Why is the "two networks" construction an important reform for forcing a revolutionary redesign of administrative services and urban governance processes? *Shangguan Observer*. Retrieved April 11, 2021, from http://www.sdx.sh.cn/html/xw/mtgz/140280.html

Zhou, X. L. (2005). New public management: Reflection, critique, and transcendence—A review of the new public service theory. *Public Administration Journal, 2*(1), 43–48.

Chapter 2
Next-Generation Information and Communication Technology and Digital Urban Governance

As early as the 1980s, the academic community began using the term information and communication technology (ICT). Since 1997, many countries, including the United Kingdom, have gradually adopted this term more widely. ICT is a new technological field formed by the integration of information technology (IT) and communication technology (CT). Traditionally, IT focuses on the management, processing, and presentation of information, such as using computers to develop various services and applications, while CT focuses on the technological implementation of information transmission and the development and application of related devices and software, such as transmission access, network switching, mobile communication, as well as devices like radios, televisions, and satellite systems. With technological advancements and iteration, these two technologies have gradually become inseparable and merged into a new field. Today, next-generation ICTs, represented by cloud computing, big data, the Internet of Things (IoT), and artificial intelligence (AI), are developing rapidly. They are strategic, fundamental, and pioneering industries that build national information infrastructure, provide network and information services, and comprehensively support economic and social development. They also provide strong support for urban governance.

This chapter introduces some representative technologies in next-generation ICT and their applications in digital urban governance. Specifically, the representative technologies include the Internet of Things (IoT), big data, cloud computing, mobile internet, artificial intelligence, and blockchain technology.

W. Zhang and Y. Zhang, *Digital Urban Governance in China*, Qizhen Humanities and Social Sciences Library, https://doi.org/10.1007/978-981-96-6325-5_2

1 The Internet of Things and Digital Urban Governance

1.1 What Is the Internet of Things?

The Internet of Things (IoT) is a network where different sensing devices exchange and communicate information according to predefined protocols, enabling intelligent identification, positioning, tracking, monitoring, and management of objects. To some extent, we can say that the IoT is the "internet of everything," which integrates various sensing devices with the internet to form a vast network. The IoT extends communication from person-to-person to person-to-object and object-to-object, thus enabling interconnectivity between people, machines, and things at any time and in any place.

The IoT has three fundamental characteristics. The first is comprehensive sensing, meaning the IoT can use various sensing devices such as radio frequency identification (RFID), quick response (QR) codes, infrared sensors, global positioning systems (GPS), and laser scanners to obtain a variety of information about objects. The second characteristic is reliable transmission, meaning the IoT can transmit information from objects in real time and accurately for information exchange and sharing. The third characteristic is intelligent processing, where the IoT uses various intelligent technologies to analyze and process data, thereby enabling intelligent monitoring and control (Gan, 2010).

Sensors, RFID, and QR codes are some of the most widely used IoT sensing devices. A sensor is a detection device that can sense the measured information and convert it into an electrical signal or other required forms of information output according to certain rules to meet a series of requirements such as information transmission, processing, storage, display, and control. For example, agricultural sensors can be used to monitor soil, water, and plant temperatures. Sensors can be divided into physical and chemical types. Physical sensors mainly rely on physical effects such as the piezoelectric effect, magnetostrictive effect, polarization, thermoelectric, photoelectric, and magnetoelectric to sense the external world. Chemical sensors, on the other hand, mainly include sensors based on chemical adsorption, electrochemical reactions, and other phenomena.

RFID can identify specific targets and read and write related data through radio signals without requiring mechanical or optical contact between the identification system and the target, making it a representative automatic identification technology. RFID does not require physical contact between both parties, so it can establish connections and communication through barriers like plastic, paper, wood, etc. Its information processing speed is fast, and the transmission process typically takes less than 100 ms. High-frequency RFID readers can identify and read multiple tags simultaneously, significantly improving the efficiency of information transmission. It also has strong environmental adaptability and resistance to interference, working in different weather conditions, unaffected by pollution, humidity, and mechanical wear. Furthermore, RFID has a simple tag structure and reading equipment, greatly enhancing its usability and popularity (Lu, 2018).

A QR code, also known as a two-dimensional barcode, consists of specific geometric shapes arranged in a regular pattern on a two-dimensional plane, with black and white alternating regions that record data information. Compared to a one-dimensional barcode, a QR code can carry more digital information and record more complex content, such as network links. Due to its rich information, low cost, strong fault tolerance, and broad coding range, QR codes are widely used in commercial activities. For example, through Alibaba's Alipay, users can conveniently make bank transfers and send red envelopes via QR codes.

1.2 The Application of the Internet of Things in Urban Governance

The IoT has broad applications in urban governance. In order to improve the level of intelligent governance at the community level, the government of the Guanyintang area in Jianggan District, Hangzhou, Zhejiang Province, led Caihe Neighborhood[1] to collaborate with an IoT technology company and created an integrated smart community platform called "Guanyintang Smart Central Brain Command Center," which includes functions such as personnel, vehicle, and property management, smart fire safety, and security. Since Hangzhou has a city-scale governance platform known as the "City Brain," this command center is referred to as the "smart central brain" at the district level. The smart central brain uses many IoT technologies to meet the needs of community residents, managers, and property management. For example, at the end of 2019, 52 new cameras were installed in the Guanyintang area. The smart monitoring system in the smart central brain can view the footage in real-time. When a vehicle is parked in a no-parking zone, the command center's screen immediately shows an alarm, and the property management can quickly dispatch patrollers to handle the situation. The garages in the three communities under Guanyintang—Jingyi, Jiangting, and Changqingyuan—are equipped with independent smoke detectors. If a fire occurs, the alarm information will also be displayed on the command center's screen, and staff can quickly arrange personnel to handle the situation. Additionally, these communities are equipped with facial recognition devices, allowing residents to enter and exit the community by scanning their faces (Lin, 2020). Due to its excellent performance in community governance, the Guanyintang Smart Central Brain was successfully included in the "Top Ten Cases of Smart Community Governance" at the 5th China (Hangzhou) Smart City and Artificial Intelligence Industry Summit held in October 2020.

[1] In 2021, the Jianggan District was abolished, and the Caihe Neighborhood was placed under the jurisdiction of Shangcheng District.

2 Big Data and Digital Urban Governance

2.1 What Is Big Data?

Big data can be understood as large and complex datasets that traditional data processing software cannot handle effectively within a specific timeframe. For example, the massive user behavior data generated by WeChat, which has 1.09 billion daily active users, is obviously impossible to process using conventional software such as Excel. As early as 2001, the world-renowned IT research and analysis firm Gartner defined the term big data. Around 2005, people began to realize that online services like Facebook and YouTube also generate vast amounts of data, providing new opportunities to understand user behavior. Subsequently, tools developed specifically for storing and analyzing large datasets, such as Hadoop, Spark, and NoSQL, significantly reduced the costs of data storage and analysis, making big data easier to store, analyze, and use. This led to an explosive growth in the volume of big data. Although the concept of big data emerged in the early twenty-first century, large datasets actually existed as early as the 1960s, with the first global data centers and the first relational databases being developed at that time (Oracle China, 2021).

Demchenko et al. (2013) proposed that big data has five key characteristics (the 5Vs):

(1) **Volume**: The primary characteristic of big data is the massive volume of data, including data collection, storage, and analysis. In real-world applications, big data often starts at the terabyte (TB) or petabyte (PB) scale.[2] In the social sciences, the data volume tends to be somewhat lower. It's important to note that the term "big data" is often misused in daily life; for example, news reports sometimes describe datasets that are only slightly larger than traditional ones as big data, such as statistical descriptions of thousands of university freshmen's departments or gender information.

(2) **Velocity**: Data collection, transmission, and processing speeds are fast, with high timeliness requirements. In some internet applications, such as searching for information through Baidu or Google search engines, real-time or near-real-time calculations are necessary to provide users with timely and accurate feedback.

(3) **Variety**: This refers to the diverse types of data, which include not only structured data that can be stored in relational databases but also a vast amount of unstructured data, such as text, images, audio, and video.

(4) **Value**: While large volumes of data inherently contain value, the value density is relatively low compared to the vast amount of data. Thus, appropriate data mining methods are needed to extract valuable information from big data.

(5) **Veracity**: Big data is closely related to the real world, so ensuring the accuracy and reliability of data is essential. Only truthful and reliable big data is meaningful.

[2] TB and PB are units of computer storage. 1 PB equals 1024 TB, and 1 TB equals 1024 GB.

Big data includes structured, semi-structured, and unstructured data. Structured data can be represented and stored in relational databases, typically in a two-dimensional form, such as passenger card swipe data, which includes fixed information such as passenger card numbers, swipe stations, and swipe times. The storage and arrangement of structured data are highly organized, facilitating queries, modifications, and analysis. Unstructured data refers to data that has an irregular or incomplete structure, making it difficult to represent using predefined data models, and unsuitable for storage in relational databases. Common examples include documents, images, audio, and video. Semi-structured data can be considered a type of structured data. While it is not suitable for storage in relational databases, it includes relevant tags to separate semantic elements and obtain information. Common semi-structured data languages include XML (Extensible Markup Language) and JSON (JavaScript Object Notation). XML is used to mark up electronic files, giving them a structured markup language, while JSON is a lightweight data exchange format known for its readability and ease of use.

2.2 *Urban Big Data Platforms*

Urban big data is data closely related to the physical and social spaces of cities, typically corresponding to the city's functional departments. Digital urban governance relies on the support of urban big data, and its main form is the construction of unified urban big data platforms covering the entire city ("city as platform"). Urban big data platforms are central hubs for the centralized storage and processing of urban big data, providing data application services. Constructing urban big data platforms can enhance cross-departmental and cross-industry coordination, improve resource integration and utilization, and help accelerate the organized aggregation, deep sharing, related analysis, and efficient use of urban information resources. These platforms provide collaborative services across levels, regions, departments, and businesses for different urban governance stakeholders (China Academy of Information and Communications Technology, 2019).

The China Academy of Information and Communications Technology conducted a survey of the e-government or urban big data platforms of 36 major cities in China, including platforms like Shanghai's data-sharing exchange platform and Hangzhou's City Brain. The results show that as of May 2019, most cities were still in the stage of developing government data sharing and exchange platforms, with only a few cities like Beijing, Shanghai, Hangzhou, and Guiyang showing significant urban management and service features, having initially built city-level big data platforms (China Academy of Information and Communications Technology, 2019). Although the big data platforms of major cities in China are still in their immature stages, their development is progressing rapidly. For example, since its launch in August 2017, the Hefei big data platform has integrated nearly 200 business systems across the city, collecting 8,840 types of data and 23.1 billion data entries. It has launched 6,198 categories of data resource sharing directories, 189 categories of electronic certificates, and 31.35 million sets, and has enabled the cross-city handling of 65 items across 41 cities in the Yangtze River Delta (Wu, 2021).

In addition, to promote the integration and use of big data resources, many cities have established dedicated big data management agencies. The Hangzhou Municipal Data Resource Management Bureau was established by the Hangzhou Municipal People's Government in January 2017 as a government department responsible for coordinating the management of data resources. Its goal is to promote the digital transformation of the government, build a modern urban governance system led by digitalization and supported by big data, foster the development of Hangzhou's big data industry, guide the construction of data security assurance systems, organize and implement the security protection of government and public data, and advance the development of big data talent. The Shanghai Big Data Center was officially established in April 2018 as a fully funded public institution under the Shanghai Municipal Government Office. It is responsible for the collection and integration of various data, including government data, industry data, and social data. The center oversees the construction and operation of the "One Network for All Services" main portal, the "SuiShenBan" super app, as well as the development and management of new infrastructure such as Shanghai's e-government cloud, government extranet, and disaster recovery center. To promote the cultivation of high-level talent in the field of big data in Shanghai, the Shanghai Big Data Center established a postdoctoral research workstation in January 2021. This is the first postdoctoral research station approved for a provincial-level big data management institution in the country. The Big Data Center will collaborate closely with universities such as Fudan University, Shanghai Jiao Tong University, and East China Normal University to jointly recruit and train postdoctoral researchers, aiming to foster high-level research and application outcomes that will comprehensively empower the modernization and iteration of urban governance through digital means.

3 Cloud Computing and Digital Urban Governance

3.1 What Is Cloud Computing?

Amazon, Microsoft, and Alibaba are the top three global cloud computing service providers. According to data from Gartner's global cloud computing market tracking, Amazon AWS holds a 47.9% market share, Microsoft Azure holds 15.6%, and Alibaba Cloud holds 7.7%. Together, these three providers account for 71.2%, far surpassing other vendors (Chen, 2020). Amazon defines cloud computing as providing computing resources on demand over the internet, with a pay-per-use pricing model. Microsoft views cloud computing as the delivery of computing services, including servers, storage, databases, networking, software, analytics, and intelligence services, via the Internet (the "cloud"), providing rapid innovation, elastic resources, and economies of scale. For cloud services, users typically pay only for what they use, which helps reduce operational costs, ensures more efficient infrastructure operation, and allows for adjustments based on changing business needs. If we liken the cloud computing industry to the electricity industry, the process is similar to "power generation—transmission—consumption": software and hardware are centrally deployed in cloud computing centers or platforms (the "power plants"), users access cloud computing resources (the "electricity consumption" process), and the internet acts as the "power line." Just as we don't need a generator in every household but buy electricity, users don't need to buy hardware and software but can directly use cloud computing resources (Jin & Guo, 2016).

Cloud computing can be divided into three main types based on service content: Infrastructure as a Service (IaaS), Platform as a Service (PaaS), and Software as a Service (SaaS). (1) Infrastructure as a Service (IaaS): This provides access to network functions, computers, and data storage devices. Users don't need to purchase hardware; they can pay to use the cloud provider's hardware. For example, we can directly use Baidu Netdisk or Alibaba Netdisk without purchasing a hard drive. (2) Platform as a Service (PaaS): This allows users to focus on application deployment and management, without managing the underlying infrastructure. Cloud service providers offer environments for software development, testing, and running, so users can open, test, and run their software on the cloud platform. (3) Software as a Service (SaaS): This provides a complete product whose operation and management are handled by the service provider. In most cases, SaaS refers to end-user applications, such as the web-based NetEase email service. Users do not need to purchase the software themselves; instead, they pay for the software's usage through a service fee and access it online.

Based on usage, cloud computing can be classified into three types: Public Cloud, Private Cloud, and Hybrid Cloud. (1) Public Cloud: Multiple users share the IT resources of a single cloud service provider. This is the most common cloud computing deployment model, with all hardware and software managed by the cloud provider, such as Microsoft Azure or Alibaba Cloud. Users pay for the IT resources they use; (2) Private Cloud: A cloud computing center or platform built by an enterprise or organization for its exclusive use, without being open to the public. The private cloud may be hosted by a third-party provider but is always maintained on a private network for the enterprise's use. This type is typically used by government agencies, large enterprises, financial institutions, or other organizations seeking greater control. (3) Hybrid Cloud: A combination of public and private clouds, where some resources are shared and accessible externally, while others are kept private. Hybrid clouds are suitable for enterprises or organizations with excess IT resources, allowing them to meet their own needs while offering excess resources to others.

Microsoft China (2021) identifies seven major advantages of cloud computing:

(1) Low Cost: Cloud computing allows users to avoid purchasing hardware and investing in setting up and operating on-site data centers. User costs are only a fraction of traditional IT costs.
(2) Timely Speed: Cloud computing services are typically offered on-demand (pay-as-you-go), allowing users to quickly allocate vast computing resources with

just a few clicks, providing businesses with great flexibility and eliminating the pressure of capacity planning.

(3) Global Scalability: Cloud computing services offer elastic scaling capabilities, allowing users to obtain the right amount of IT resources, such as additional or less computing power, storage space, and bandwidth, based on needs.
(4) Work Efficiency: On-site data centers often require extensive hardware setup, software patches, and other time-consuming IT management tasks. Cloud computing can eliminate most of these tasks, enabling users to focus on more important business objectives.
(5) High Performance: Cloud computing services run on secure data centers across a global network, with regular upgrades to fast, efficient computing hardware. Compared to a single enterprise data center, it offers many benefits, including reduced application network latency and improved scalability economics.
(6) Reliability: By mirroring data across multiple redundant sites in the cloud provider's network, cloud computing simplifies data backup, disaster recovery, and business continuity at a lower cost.
(7) Security: Cloud computing provides strategies, technologies, and controls that enhance overall security, helping protect data, applications, and infrastructure from potential threats.

3.2 The Application of Cloud Computing in Urban Governance

Cloud computing technology aligns well with the current trend of centralized electronic governance and digital administration in some cities in China. For example, cities are building cloud-based data centers or supercomputing centers, promoting the centralization of government departments' data centers to achieve unified operation and maintenance. Cloud computing technology is also being used to build a network of city government websites, with a main portal site for the city's government and subsidiary websites for various departments. Additionally, cloud-based integrated information service platforms are being developed to promote interoperability of business applications, facilitate information sharing, and enhance business collaboration. In December 2009, the Chengdu Cloud Computing Center was launched as China's first city-level cloud computing center, built through a model of enterprise investment, operation, and management with government procurement of services. This initiative pioneered a transformation in the construction and management model of cloud computing centers in China: the government shifted from funding the "center" to paying for "services," while enterprises moved from being equipment suppliers to cloud service providers. The Chengdu Cloud Computing Center has connected over 300 application systems from more than 100 departments of the Chengdu municipal government and its district and county governments. One prominent example of a cloud service application is the "Tianfu Citizen Cloud." The "Tianfu Citizen Cloud" is a one-stop, cross-platform urban service platform driven by citizens' needs, which consolidates services from various Chengdu city departments. It serves as a window for the government to actively respond to public demands. After more than a year since its launch, "Tianfu Citizen Cloud" has accumulated over 3 million registered users and offers more than 180 online services covering the full life cycle, including services related to infants, education, livelihood, transportation, and elderly care. Users can access comprehensive civic services, such as checking information, making appointments, handling administrative matters, and paying fees, all through a single, real-name authenticated account (Yu, 2020a).

4 Mobile Internet and Digital Urban Governance

4.1 What Is Mobile Internet?

Mobile Internet is the product of combining Internet technologies, platforms, business models, and mobile communication technologies. With the development of mobile communication from 3G, 4G, to 5G, as well as the widespread use of smartphones, tablets, and other mobile terminals, we have entered the era of Mobile Internet. Now, users can conveniently access the Internet anytime and anywhere to access a wealth of resources and services. According to *China's Internet Development Report (2020)* compiled by the Internet Society of China, by the end of 2019, the number of mobile Internet users in China reached 1.319 billion, accounting for 32.17% of the global total; the number of 4G base stations totaled 5.44 million, more than half of the global total. China's development of mobile Internet is at the forefront of the world (Tang, 2020a).

Smartphones are the most significant mobile terminal devices. They feature independent mobile operating systems, allowing users to install applications for browsing websites, playing games, socializing, and more. Therefore, smartphones not only perform basic functions like making calls but also enable users to browse the web, listen to music, watch videos, use navigation tools, and engage in social interactions. According to *China's Mobile Internet Development Report (2020)*, as of March 2020, there were 897 million mobile internet users in China, and in 2019, the total shipment of mobile phones reached 389 million units, including 13.769 million 5G phones (Tang, 2020b). Furthermore, mobile terminals like tablets, smartwatches, and in-car devices have shown rapid development. For instance, in 2019, the shipment of wearable devices in China reached 99.24 million units, an increase of 37.1%.

Here, we explain a mobile Internet term closely related to digital urban governance, known as "WWAD," referring to Weibo, WeChat, mobile apps, and Douyin. This term neatly summarizes the most popular mobile applications and the trend of mobilization in China. These mobile media significantly enhance digital urban governance capabilities. It is worth noting that in different contexts, variations like "WWA" and "WWADH"[3] may also appear.

(1) Weibo: Weibo is a social media platform based on user relationships, enabling the sharing, dissemination, and retrieval of information instantly. Sina launched "Sina Weibo" in August 2009, making it the first major portal website in China to offer microblogging services.
(2) WeChat: WeChat is an instant messaging application developed by Tencent, launched in January 2011 for smart devices. WeChat supports a range of functions, including sending videos, images, text, and voice messages, creating group chats of up to 500 people, real-time video chats with up to 9 people, and sharing life moments with friends.
(3) Mobile Apps: Mobile apps are applications primarily installed on smart mobile devices. The spread of the mobile Internet has given rise to many mobile apps, including Weibo, WeChat, and Douyin, each operating as standalone apps. *China's Mobile Internet Development Report (2020)* indicates that, by 2019, there were 3.67 million applications in China, with the top four categories being games, daily tools, e-commerce, and life services, which together accounted for 57.9% of the total app quantity.
(4) Douyin: Launched by ByteDance in September 2016, Douyin is a creative short video and music-based social app. Users can select songs and create short videos, with Douyin recommending videos based on the users' preferences. Currently, Douyin, along with its international version TikTok, has become the most popular short video platform globally, regularly topping app download charts.

4.2 *The Application of Mobile Internet in Urban Governance*

A typical mobile Internet application related to digital urban governance is Shanghai Release. Shanghai Release is a government media account operated by the Shanghai Municipal Government's Information Office on platforms such as Weibo, WeChat, Toutiao, and Douyin. It was the first city-level government official Weibo account in China. In November 2011, Shanghai became the first city to open a Weibo account called Shanghai Release, and in June 2013, it launched its WeChat official account. Shanghai Release is the most influential city government account in China. As of November 2019, Shanghai Release's Douyin account had accumulated over 300 million views, received more than 10 million likes, and gained over 880,000 followers. The most viewed single short video had over 30 million views, 2.31 million likes, 60,000 comments, and 180,000 shares. During the 2nd China International Import Expo, Shanghai Release published 41 short videos on Douyin, which collectively garnered over 24 million views (Lu, 2019). By October 2020, the number of followers on the WeChat official account of Shanghai Release surpassed

[3] H in "WWADH" refers to applications such as Toutiao's "Toutiao Hao" and Baidu's "Baijia Hao."

8 million, and by the same year, over 130 articles had received more than 1 million reads, with a single article reaching nearly 19 million reads (Pengpai News, 2020).

The content of Shanghai Release primarily consists of general information covering various topics such as policies, weather, transportation, education, culture, and healthcare in Shanghai. The primary audience consists of Shanghai residents, providing them with convenient information services for daily life. For example, on the WeChat official account of Shanghai Release, there is a "Municipal Services Hall" where citizens can access a variety of services, including real-time bus arrival information, traffic violations, road conditions, school enrollment details, garbage classification queries, housing fund inquiries, social security card applications, hospital report checks, and name queries. Under the guidance of the Shanghai Big Data Center, Shanghai Release has also integrated the "Suishenban" WeChat mini-program, further expanding the online government service channels for citizens.

Shanghai Release has several distinct features:

(1) Cross-media platform operation, diverse content forms: Shanghai Release fully utilizes new media platforms commonly used by citizens in the mobile Internet era. From Weibo short posts to WeChat long articles and Douyin short videos, it ensures that citizens with different preferences can access content through suitable channels. When releasing information, Shanghai Release emphasizes the use of multimedia, including images, text, and video, while paying attention to page layout and formatting to improve readability and overall visual comfort.
(2) Authoritative yet approachable: Shanghai Release represents the virtual image of the Shanghai Municipal Government Information Office and must align with the tone of government agencies to ensure the accuracy and authority of major policies and government information. At the same time, Shanghai Release often engages with the daily lives of ordinary citizens in a relaxed and fun manner, opening up channels for governance, public participation, and political discussion. It has become a platform that the public relies on and trusts. Shanghai Release encourages user-generated content through methods such as collecting opinions, soliciting names, and conducting online surveys, fostering positive interaction between government agencies and citizens. For instance, Shanghai Release solicited name suggestions for its mascot, which was selected via public voting. The mascot was ultimately chosen to be a small rabbit with a round face, named "Bunny Xiaobu." The design is cute and approachable, and the name is inspired by the famous Shanghai snack, the White Rabbit Milk Candy. The design style also draws on another well-loved Shanghai treat, the Xiaolongbao (soup dumplings), making it a favorite among locals.
(3) Communication and service functions: In addition to disseminating government information, Shanghai Release serves as a bridge between the government and citizens. For example, its WeChat account brings government services to mobile devices, offering two primary services: "Municipal Services Hall" and "Suishenban." These provide a variety of online urban services, allowing users to handle various public affairs from their mobile devices, making it easier to

take care of some simple civil matters without leaving home (Li, 2019; Liu, 2017).

5 Artificial Intelligence and Digital Urban Governance

5.1 What Is Artificial Intelligence?

Artificial Intelligence (AI) is the science that studies theories, methods, technologies, and their applications to simulate, extend, and enhance human intelligence. AI has the ability to interpret external data, learn from it, and utilize the acquired knowledge to achieve specific goals and tasks. The core challenge of AI lies in constructing abilities that are similar to or surpass human reasoning, planning, learning, perception, and tool use, which enables intelligent machines to hear (e.g., speech recognition), see (e.g., image recognition), speak (e.g., human-computer interaction), think (e.g., human-computer chess), and act (e.g., autonomous vehicles). The primary fields of AI research include deep learning, intelligent robotics, image recognition, computer vision, natural language processing, and expert systems.

The field of AI was formally established during the "Summer Research Project on Artificial Intelligence" held at Dartmouth College in 1956. Over the following decades, AI research encountered several periods of stagnation, as researchers significantly underestimated the difficulty of achieving AI, leading to governments and corporations withdrawing funding when returns were not forthcoming. After several ups and downs, AI made notable progress. In the 1990s, the development of network technologies, particularly the internet, accelerated AI research, pushing AI technology toward practical applications. A landmark event occurred in May 1997 when IBM's "Deep Blue" computer program defeated the world chess champion Garry Kasparov. Entering the twenty-first century, advancements in information technologies such as big data, cloud computing, and the Internet of Things fueled the rapid development of AI, bridging the technological gap between science and application. One such breakthrough was the AlphaGo program, developed by the DeepMind team under Google, designed for the game of Go. In March 2016, AlphaGo defeated world Go champion Lee Sedol with a 4-1 score. Unlike human intelligence, AlphaGo improved its skills through deep learning by inputting millions of games played by top human players. Beyond human-computer chess, AI technologies have now made breakthroughs in areas such as image classification, speech recognition, question answering, and autonomous vehicles, evolving from "not usable or difficult to use" to "usable."

Due to its uniqueness, many countries have developed national AI strategies. In October 2016, the United States released the *National Artificial Intelligence Research and Development Strategic Plan*, outlining an overall framework and seven priority strategies for leveraging AI to enhance economic power and improve social security. In February 2020, the European Commission released the "White Paper on Artificial Intelligence" in Brussels, aiming to foster innovation in AI and promote the development of ethical and trustworthy AI. In November 2019, Singapore launched an 11-year national AI strategy, laying out a vision, methods, key plans, and AI ecosystem development for Singapore's future. The strategy focuses on nine industries or areas, including transportation and logistics, manufacturing, finance, security, cybersecurity, smart cities and real estate, healthcare, education, and government. We can see that many aspects of Singapore's AI strategy are closely tied to digital urban governance. In July 2017, the State Council of China issued the "Next Generation Artificial Intelligence Development Plan," outlining a three-step strategy: (1) By 2020, AI technology and applications will be on par with the world's advanced levels, with AI becoming a new important economic growth point and a significant path to improving people's lives, supporting the goal of building an innovative country and achieving a well-off society. (2) By 2025, AI basic theories will achieve major breakthroughs, some technologies and applications will be world-leading, and AI will become a major driving force for industry upgrades and economic transformation. (3) By 2030, AI theory, technology, and applications will be at the world's leading level, establishing China as a key AI innovation hub and significantly contributing to the smart economy and society.

5.2 The Application of Artificial Intelligence in Urban Governance

Facial recognition is a popular area of AI research that plays a crucial role in digital urban governance. It is a biometric technology that identifies individuals based on facial features, analyzing and comparing the visual characteristics of faces for identification. Broadly, facial recognition encompasses technologies like facial image capture, facial location, facial recognition preprocessing, identity confirmation, and identity searching.

Facial recognition can sometimes replace identification documents like IDs for identity verification and supports simultaneous detection of multiple individuals. The process doesn't require active cooperation from the individual, is inconspicuous, and doesn't require physical contact; as long as a face is captured, identification can take place. Due to these advantages, facial recognition has wide applications in digital urban governance, including at airports for facial recognition channels, face-based access to communities, and monitoring pedestrian traffic violations such as running red lights.

In August 2020, Xinyi City in Jiangsu Province began using facial recognition technology to automatically capture violations by pedestrians and non-motorized vehicles within its main urban area. When traffic violations such as running red lights, illegal turns, or parking over the line occur, an electronic surveillance system automatically tracks and captures images of the violators' faces. Using AI-driven identification technology, the system matches the facial information with a database to confirm the violator's identity. The captured videos and photos of the violation are displayed on roadside screens, and violators are required to attend traffic rule education sessions and face penalties. Repeat offenders who refuse to comply will have their violation records included in the credit system (Xinyi Portal, 2021). Another application of facial recognition in urban settings is the use of "face-based entry" for train stations, introduced in the 2017 Spring Festival travel season. Automatic ticket gates equipped with cameras capture passengers' facial information when they approach, comparing it to the photo on their ID cards. If the information matches, the gate automatically opens. In October 2020, Wuhan Tianhe International Airport also launched a "face-based boarding" self-service feature. Passengers can check in without printing boarding passes, simply by scanning their ID and using facial recognition, enabling easy passage through security and boarding gates. On average, it takes 3–5 s for a passenger to pass, and using self-service devices throughout the journey saves 20–30 min of queuing time (Shi, 2021).

6 Blockchain and Digital Urban Governance

6.1 What Is Blockchain?

Blockchain originated from Bitcoin. In November 2008, a person using the pseudonym Satoshi Nakamoto published the paper "Bitcoin: A Peer-to-Peer Electronic Cash System," which outlined the architecture of an electronic cash system based on P2P (peer-to-peer) network technology, encryption technology, timestamp technology, and blockchain technology. This marked the birth of Bitcoin. Although different countries have complex attitudes toward Bitcoin due to financial risks, blockchain—the core technology behind Bitcoin—is increasingly being valued. Blockchain provides a decentralized trust-building mechanism and is considered a disruptive technology. Blockchain establishes a distributed ledger database on a peer-to-peer network, creating a database jointly maintained by multiple parties, which cannot be tampered with by any individual. This database can record all historical transaction information and related data, which are distributed across various locations and are open and transparent, rather than being stored in a central server (Zhang & Wen, 2017).

As a distributed database, blockchain technology features decentralization, the removal of intermediaries, openness, reliability, and permanence. Blockchain's core technology consists of three fundamentals:

(1) Transaction: It refers to a change that occurs within the database network, such as a transfer, an event notification, or a piece of information.
(2) Block: After accumulating a certain number of transactions, a settlement occurs, and the accumulated transaction information is encrypted into a transaction block.
(3) Chain: The blocks are linked in chronological order of settlement, forming a chain of data that extends continuously, similar to a lock-chain (Singtel, 2021).

Blockchain generally includes three types: public, private, and consortium blockchains (Zhang, 2016). A public blockchain is one that anyone can join and participate in; the most typical application is Bitcoin. Public blockchains are the earliest and most widely used blockchain type. Their drawbacks include the need for significant computational power, low or no transaction privacy, and weaker security. Similar to public blockchain networks, private blockchains are also decentralized peer-to-peer networks, but the entire network is managed by a company, organization, or individual. The organization controls who can participate in the network, enforces consensus protocols, and maintains a shared ledger. Consortium blockchains designate multiple pre-selected nodes within a group to act as ledger keepers, and the creation of each block is decided collectively by these nodes. Other connecting nodes can participate in transactions but do not engage in the ledger-keeping process; anyone can access the blockchain's open API (Application Programming Interface) for limited queries.

In December 2013, the People's Bank of China, the Ministry of Industry and Information Technology, the China Banking Regulatory Commission, the China Securities Regulatory Commission, and the China Insurance Regulatory Commission jointly issued the "Notice on Preventing Bitcoin Risks." This notice emphasized that Bitcoin is not a currency, but only a virtual commodity, and that no financial or payment institution should price products or services in Bitcoin. However, the Chinese government has placed significant importance on blockchain, the underlying technology of Bitcoin. In 2014, the People's Bank of China established a special research team for the study of legal digital currency to explore the feasibility of central bank-issued legal digital currencies. In October 2019, President Xi Jinping, during the 18th collective study of the Political Bureau of the Communist Party of China, highlighted the need to treat blockchain as a core technological breakthrough for independent innovation. He called for strengthening blockchain standardization research, enhancing international discourse influence, promoting the deep integration of blockchain with the real economy, and implementing blockchain governance under the rule of law to ensure its safe and orderly development.[4] Starting in April 2020, pilot programs for central bank digital currencies began in cities like Shenzhen, Chengdu, and Suzhou.

6.2 The Application of Blockchain in Urban Governance

One of the newer applications of blockchain technology in digital urban governance is targeted poverty alleviation. Targeted poverty alleviation is an approach to poverty reduction that uses scientific and effective procedures to precisely identify, assist, and manage impoverished households based on the conditions of specific regions

---

[4] Making Blockchain Technology a Core Focus for Independent Innovation and Key Breakthroughs: Accelerating the Development of Blockchain Technology and Industrial Innovation. *People's Daily*, October 26, 2019 (1).

and households. Its core concept is "targeted," reflecting a pursuit of social fairness. In targeted poverty alleviation, blockchain technology has at least three advantages. (1) Transparency in the use of funds: Every approval of poverty alleviation funds is recorded on the blockchain, with each link in the process holding clear accountability. Government departments and banking institutions automatically join the supervision process, ensuring transparency in the entire approval process and minimizing the potential for corruption. (2) Accurate allocation of funds: Traditional funding requires multiple steps of allocation before determining the specific projects. Blockchain technology allows the determination of funding projects and usage first, followed by the appropriate allocation of funds, transforming the approach from "blanket distribution" to "precise targeting." (3) Efficient fund management: The cross-chain integration of financial services and poverty alleviation fund administrative approval chains allows for a seamless combination of blockchain and big data technology. This enables governments at all levels to track in real-time the demand, allocation, disbursement, and actual use of poverty alleviation funds (Shi, 2017).

Many provinces have already begun implementing blockchain-based targeted poverty alleviation practices. In October 2018, the Industrial and Commercial Bank of China, in collaboration with the Guizhou Guimin Group, used cross-chain integration between the banking financial service chain and government poverty alleviation fund administrative approval chain to ensure the transparent use, accurate allocation, and efficient management of poverty alleviation funds, successfully distributing the first batch of funds—1.57 million RMB (Shi, 2017). China Life Insurance has attempted to use blockchain technology and insurance mechanisms to support social poverty alleviation, focusing on the health and medical security of the main labor force in impoverished households, offering a dedicated insurance program for registered impoverished households aged 18 to 60. Through a blockchain framework, insurance data is connected in real time, and all parties involved—donors, public organizations, insurance companies, and insured individuals—participate in a joint ledger, making it impossible for any party to tamper with the records (Yu, 2020b).

References

Chen, T. M. (2020, April 23). Amazon's global cloud market share drops to 45%: Microsoft and Alibaba expand their shares. *The Paper*. https://www.thepaper.cn/newsDetail_forward_7100270

China Academy of Information and Communications Technology. (2019, March 12). *White Paper on Urban Big Data Platforms*. http://www.caict.ac.cn/kxyj/qwfb/bps/201906/P020190604489340368704.pdf

Demchenko, Y., Grossop, T., De Laat, C., et al. (2013, March 4). Addressing big data issues in scientific data infrastructure. *Proceedings of the 2013 International Conference on Collaboration Technologies and Systems*. http://atlas.ch/.10.1109/CTS.2013.6567203

Gan, Z. X. (2010). A study on the origin and development of the Internet of Things. *Modern Economic Information, 1*, 158–157.

Jin, J. J., & Guo, Y. L. (2016). *Smart cities: Big data and urban governance in the internet age*. Electronic Industry Press.

Lin, J. A. (2020, April 9). Where the traffic jams are, where the problems lie: "Guanyintang Smart Middle Brain" knows it all. *Urban Express*. https://new.qq.com/omn/20200409/20200409A0ELW700.html

Li, Q. (2019). A study on the content production logic of urban government WeChat official accounts: A case study of "Shanghai Release." *Journalism and Writing, 10*, 94–98.

Liu, W. (2017). *A study on the WeChat official account "Shanghai Release" in urban government affairs* (Unpublished master's thesis). Xiangtan University, Xiangtan, China.

Lu, T. Y. (2019, November 28). Available on 13 platforms with a total fan base of over 14 million: "Shanghai Release" maintains its leading position among national government new media. *Why*. http://www.why.com.cn/wx/article/2019/11/28/15749358071181918970.html

Lu, X. B. (2018). An analysis of radio frequency identification technology. *China New Communications, 1*, 67–68.

Pengpai News (2020). *Shanghai Release official WeChat account surpasses 8 million followers*. https://www.163.com/dy/article/FOE1LNI80514EGPO.html.

Microsoft China. (2021, March 3). *What is cloud computing? A beginner's guide*. https://azure.microsoft.com/zh-cn/overview/what-is-cloud-computing/

Oracle China. (2021, March 4). *What is big data?* https://www.oracle.com/cn/big-data/what-is-big-data/

Shi, H. P. (2017, October 10). Industrial and Commercial Bank of China innovates poverty alleviation financial services using blockchain technology. *Xinhua News*. http://www.xinhuanet.com/money/2017-10/10/c_129718001.htm

Shi, Q. (2021, March 11). Face recognition boarding arrives! No boarding pass required; using full self-service equipment can save nearly half an hour. *Hubei News*. http://news.cnhubei.com/content/2021-03/11/content_13671393.html

Singtel. (2021, March 5). *What is blockchain?* https://www.singtel.com/business/articles/what-is-blockchain

Tang, Q. (2020a, July 23). The scale of mobile internet users in China exceeds 1.3 billion, accounting for more than 30% of the global internet user base. *The Paper*. https://www.thepaper.cn/newsDetail_forward_8405910

Tang, W. H. (2020b, July 16). The "2020 China Mobile Internet Development Report" is officially released. *People's Daily*. http://yuqing.people.com.cn/n1/2020/0716/c209043-31785567.html

The Paper. (2020, October 8). *"Shanghai Release" government WeChat account surpasses 8 million followers, maintaining its leading position nationwide*. https://baijiahao.baidu.com/s?id=1679961769987198331&wfr=spider&for=pc

Wu, L. (2021, January 5). Hefei, Anhui, fully builds the "city brain," greatly reducing paper application materials. *China News Network*. http://www.ah.chinanews.com/news/2021/0105/271435.shtml

Xinyi Portal. (2021, March 12). *New Xinyi traffic police facial recognition technology for capturing traffic violations officially launched*. https://wemp.app/posts/98becc23-6e63-47b2-901b-a2c366f4ef75.

Yu, H. Q. (2020a, May 4). The first decade of the domestic "city cloud": From building centers to purchasing services, Chengdu was the first to "eat the crab." *The Paper*. https://www.thepaper.cn/newsDetail_forward_7074323

Yu, Y. (2020b, April 8). New path for poverty alleviation and helping farmers: China Life's blockchain boosts targeted poverty alleviation. *Xinhua News*. http://www.xinhuanet.com/fortune/2020-04/08/c_1125828164.htm

Zhang, J. (2016). *Blockchain: Defining the future financial and economic landscape*. Mechanical Industry Press.

Zhang, Y., & Wen, J. (2017). The IoT electric business model: Using blockchain technology for the internet of things. *Peer-to-Peer Networking and Applications, 10*(4), 983–994.

Chapter 3
New Types of Urban Data and Digital Urban Governance

Traditional urban data, such as questionnaire data, resident interviews, and statistical information, suffer from high collection costs, low efficiency, limited population coverage, and poor data accessibility. With the rapid development of next-generation information and communication technologies, a large number of new types of urban data have been generated, which better address the limitations of traditional urban data. When the devices and systems for data collection are already deployed, the cost of data collection is relatively low, and the speed is very fast, even reaching the point of real-time data collection. New types of urban data, such as mobile signaling big data and public transportation card data, can cover populations of tens of millions and include residents living and working in different areas. For the data owners, these data can be directly used to improve production and service efficiency. When the data meets the requirements for openness, it can also be made available to third-party users. It is important to note that new types of urban data not only include urban big data but also include city data that do not meet the big data definition but are difficult or impossible to obtain through traditional methods. For example, taxi trajectory data from a city is a typical example of urban big data, while restaurant data from a city's public review website is traditional data that is difficult to obtain. Both types fall under the new types of urban data described in this chapter.

In general, new types of urban data can quantitatively depict a city's physical and social space from different perspectives, helping us understand urban systems and citizen behavior patterns, thereby supporting the practice of digital urban governance. This chapter introduces several common types of new urban data and their applications in urban governance practice or research, including five major categories: smart card data, smartphone data, point-of-interest data, social media data, and trajectory data.

W. Zhang and Y. Zhang, *Digital Urban Governance in China*, Qizhen Humanities and Social Sciences Library, https://doi.org/10.1007/978-981-96-6325-5_3

1 Smart Card Data

1.1 Introduction

A smart card is a portable card with an embedded chip. Based on the chip and communication method, smart cards can be divided into storage cards, microprocessor cards, contact cards, contactless cards, and dual-interface cards. Smart cards are widely used in various social sectors such as government, banking, education, and transportation to store information such as money, fingerprints, photos, and identity IDs, and are used for identity recognition, security authentication, and payment purposes. Common examples of smart cards include bank cards, transportation cards, student cards, employee cards, and various membership cards used in different daily scenarios.

Generally, smart cards have the following advantages: (1) they are easy to carry and can be conveniently stored in a wallet or card holder; (2) they do not require physical contact with the card reader, preventing various malfunctions or damage from contact and making them usable in environments with dust and pollution; (3) they are secure, with the reader having a physical interface that is not directly accessible to the end user to ensure safety; (4) they can serve multiple functions, such as some city transport cards that are also usable for shopping in malls; (5) due to technological advancements, the manufacturing cost of smart cards is relatively low and continues to decrease each year.

The process of using a smart card generates data from the card swipe. The data structure of card swipe records is relatively simple. For example, when a passenger uses a metro card to enter a subway station, a raw smart card swipe record is generated. This record contains information such as the card number (ID), card type (e.g., regular card, senior card, student card), metro station name, metro line number, and swipe time. After processing the raw card swipe records, we can obtain more detailed passenger travel information. For example, processed metro swipe data not only includes the card number and type but also the names of the entry/exit metro stations, entry/exit times, and fare amounts. Table 1 shows the raw data for Shanghai metro card swipes, mainly containing anonymous card numbers, time, station names, fare amounts, etc. It is easy to infer that, in consecutive swipe records with the same card number, an amount of 0 corresponds to entry data, while an amount greater than 0 corresponds to exit data. Based on this, we can identify a complete travel record for each passenger based on their card number (see Table 1).

Depending on the use scenario (or card type), we can obtain different types of smart card swipe big data. For example, when using a park membership card to enter and exit the park, data related to the activity of "visiting the park" is generated. Currently, the most widely used smart card swipe data in practical research and applications mainly includes public transport cards and bank card swipe data, with a small amount of research starting to focus on student card swipe data.

Table 1 Examples of Metro card transactions Big Data in Shanghai

Anonymized card no.	Date	Time	Station name	Fees
22019252167	2016/7/16	8:32:05	Loushanguan Road (Line 2)	0
22019252167	2016/7/16	7:51:54	Shangda Road (Line 7)	3.6
22019252167	2016/7/16	16:28:42	Loushanguan Road (Line 2)	0
22019252167	2016/7/16	17:11:49	Changzhong Road (Line 7)	3.6
31042664781	2016/7/16	12:06:37	Fengqiao Road (Line 11)	0
31042664781	2016/7/16	12:31:24	Caohejing Hi-Tech Park (Line 9)	3
4002346948	2016/7/16	10:01:02	North Jiading (Line 11)	0
4002346948	2016/7/16	12:37:10	Zhongshan Park (Line 3)	5
4002346948	2016/7/16	16:02:39	Zhongshan Park (Line 2)	0
4002346948	2016/7/16	17:03:14	North Jiading (Line 11)	6

1.2 Research and Application

Excess Commuting. Excess commuting is a common transportation phenomenon in cities, referring to the deviation between actual commuting costs and the theoretically optimal commuting costs. From a city-wide perspective, excess commuting is inevitable because it is not feasible for every resident to live in the most optimal location, leading to commuting optimization challenges. However, when the deviation becomes too large, it indicates that residents are wasting excessive time on commuting, pointing to an inefficient urban spatial structure and creating a job-housing imbalance. Zhang et al. (2021) conducted a quantitative analysis of excess commuting using travel data from the Shanghai Metro in April 2015, covering 123 million metro travel records involving 10.5 million cardholders. These records helped identify 1.82 million commuters, along with their workplace and residential locations, both represented by corresponding metro stations. The results revealed that the average commuter spends 34.60 min on one-way trips, with a total of 69.20 min considering the return journey. Among all commuters, 44.5% commute five days a week, which aligns with the typical five-day workweek. However, there were other commuting patterns: 23.3% commute four days a week, and 11.4% commute six days a week, reflecting some flexible work schedules and the need for weekend overtime. Using a genetic algorithm-based model for excess commuting, we optimized Shanghai's commuting pattern by assuming that workers' locations were fixed while their residences could be adjusted to minimize commuting distances. The model sought to optimize both commuting efficiency and fairness. The results showed that, after optimization, the average commute time could be reduced from 34.60 min to 13.80 min, greatly enhancing both efficiency and fairness.

Research on Personal Privacy Issues Using Bank Card Data. In recent years, the explosion of big data has the potential to fundamentally reshape how we understand, govern, and study cities. However, data openness brings inherent privacy risks. De Montjoye et al. (2015) analyzed credit card transaction data covering 1.1 million people over a period of three months to investigate whether spatiotemporal location information in such data can compromise personal privacy. Specifically, they explored how much spatiotemporal data is needed to accurately identify an individual. They provided an example to illustrate this issue. Suppose we try to find a person named Scott in an anonymous credit card transaction dataset. The database includes only anonymous IDs, without real names or other identifying information, and each transaction record contains basic details like time, location (store), and transaction amount. Suppose we know two spatiotemporal details about Scott: he visited a bakery on September 23 and a restaurant on September 24. Upon querying the database, we find that only one person matches these two locations on these two dates, thereby identifying him as Scott. With the anonymous ID, we can retrieve all of his transaction records, such as purchases made at a store on September 23, including the amount spent. The analysis showed that knowing just four spatiotemporal data points could identify approximately 90% of individuals, and including the transaction amount could increase the identification rate by 22%. Furthermore, reducing spatiotemporal resolution (e.g., recording just the day rather than the specific time of each transaction) does not significantly improve privacy protection, and women were found to be at higher risk of identification than men.

Using Student Card Data to Understand Library Visits. Zhang and Liu (2019) analyzed student card data from a university in Chengdu to study student visits to the main library. In this study, each visit to the library was treated as an activity that could involve borrowing/returning books, studying, or participating in work-study programs. The activity was represented as a point on a two-dimensional time plane, with the start and end times of the activity as the horizontal axis. This approach is similar to how geographical information systems (GIS) use latitude and longitude to locate an object in space. By adopting this framework, the authors applied spatial point pattern analysis methods commonly used in GIS to analyze time-based activity patterns. The original dataset covered card usage from November 3 to November 30, 2014, over four weeks. After data preprocessing, 57,180 activity records from 7,412 student cards were obtained, with each activity averaging 129 min. The authors then employed spatial analysis tools like mean center, kernel density, nearest neighbor distance, and hotspot analysis to examine student behavior and patterns. They found that academically successful students were more likely to visit the library in the evening, possibly because such students preferred studying at the library during nighttime hours.

2 Smartphone Data

2.1 Introduction

Smartphones are mobile devices equipped with an independent operating system that can be used for communication and to install various applications (such as social media, shopping, navigation, and gaming). As the functionalities of smartphones have become more comprehensive and sophisticated, they have become the most frequently used smart devices in daily life. The phrase "one phone per person" aptly reflects the widespread use of smartphones.

The process of using smartphones generates a large amount of mobile phone data. This data generally falls into two categories: mobile call detail records (MCDR) and mobile signaling data. Mobile call data includes records of calls made between users, with core features such as the anonymous phone numbers of both the caller and the recipient. It also contains some personal attributes of the user. Although mobile call data can effectively capture information about a user's social network, it raises privacy concerns, which means it is rarely used in practice. This data is retained by telecommunications operators for extended periods and is generally not made available to third parties.

Mobile signaling data is captured and recorded by the telecom operators' base stations (such as China Mobile, China Unicom, and China Telecom) when users make calls, send text messages, or switch base stations. A single signal record includes information such as the user ID, timestamp, base station location code, and the type of event (such as making or receiving a call, sending or receiving a message, or updating location). Mobile signaling data is the most widely used mobile phone data in research and practice, such as supporting the understanding of urban segregation done by the authors (Zhang et al., 2024, 2025).

Compared to other new urban data types, mobile data covers the broadest range of dimensions, including population, time, and space. Aside from certain demographic groups—such as individuals under 18, those over 60, or people in economically disadvantaged areas—almost everyone owns and uses a smartphone, resulting in a very high penetration rate. According to the *Global Mobile Market Report* released in September 2020, the number of global smartphone users is expected to reach 3.5 billion, with users from China accounting for more than a quarter (Newzoo, 2020). From a temporal perspective, as soon as users begin using their smartphones, their data starts being collected, meaning the coverage period for mobile data can span several years or even decades. From a spatial perspective, because telecom operators have established a dense network of mobile signal base stations nationwide, the usage behavior of mobile users—whether in their hometown or when traveling to other provinces or countries—can be recorded. Therefore, mobile data covers a very large spatial range, including cities, regions, provinces, and even countries and the entire world.

However, privacy issues must be considered when using mobile data, especially since smartphones are registered with real names, meaning that mobile data may include a large amount of personal privacy information, such as names, ID numbers, gender, and usual residences. This sensitive information can easily lead to privacy breaches, so enhanced algorithmic anonymization is necessary. Data anonymization involves not only encrypting certain fields of personal data but also avoiding the leakage of information through the relationships between data points, such as tracking the user's movement trajectory. In the analytical process, care should be taken to avoid performing analyses that might easily expose user privacy.

Additionally, while mobile data can cover a very large spatial area, the spatial accuracy of recorded phone usage behavior is influenced by the density of mobile signal base stations. The location information in mobile data is not the user's precise location but rather the location of the mobile base station nearest to the user. This is a limitation of mobile data—for example, we cannot use mobile data to pinpoint a user's exact workplace or residence. In densely populated areas (such as city centers), mobile base stations are more densely distributed, and the distance between them is shorter, which leads to higher spatial accuracy in mobile data. In sparsely populated areas (such as rural regions), there are fewer base stations and greater distances between them, leading to lower spatial accuracy in mobile data.

Major domestic telecom operators are conducting research and consulting based on big data from smartphones. China Unicom's "Smart Footprint" is a prime example. Smart Footprint is a specialized big data company jointly established by China Unicom and Telefónica, Spain's telecommunications giant. It is the first company among domestic telecom operators to specialize in big data services. Leveraging China Unicom's extensive communication network and market experience, along with Telefónica's mature technology in big data services, Smart Footprint provides insights based on location data and related big data services (NetEase Technology, 2017). Smart Footprint possesses vast amounts of big data from smartphones, including trajectory signal data, internet behavior logs, user attribute data, and mobile terminal data. It comprehensively tracks spatial and temporal characteristics of urban residents' dwellings (such as home and work locations), travel patterns, and preference attributes. It covers over 400 million users of China Unicom, processing 160 billion location records daily, 538 billion internet activity logs, and 33 billion call records. These data are widely applied in urban planning and governance. For example, they can be used to analyze urban populations (residential, working, and visitor populations), their spatial distribution, and changes over time; track the movement dynamics of people within and across cities, and assess traffic efficiency based on residents' travel patterns to facilitate better route and transportation planning; and analyze the demographic characteristics and preferences of users of key urban infrastructure to provide data support for unified government planning (Smart Footprint, 2021).

2.2 Research and Application

COVID-19 Pandemic Research Based on Big Data from Smartphones. Jia et al. (2020) studied anonymous mobile data from over 11 million people provided by a large state-owned telecom operator in China. These individuals came from 296 counties across all 31 provinces of China but had stayed in Wuhan for at least two hours between January 1 and 24, 2020. The study found that quarantine and isolation measures were highly effective in significantly reducing population movement: from January 22 to 23, the population outflow dropped by 52%, and by January 24, it had decreased by 94%. Additionally, based on the distribution of population outflow, it was possible to accurately predict the frequency and geographic location of COVID-19 infections in China up to two weeks in advance and to identify cities with a higher potential risk of transmission in the early stages of the outbreak.

COVID-19 Pandemic Control Through Mobile Apps. In March 2020, the Singaporean government launched an app called *TraceTogether*, developed jointly by the Government Technology Agency and the Ministry of Health. This app, designed for COVID-19 pandemic prevention and control, allows phones to exchange short-range Bluetooth signals to record individuals with whom the user has been in close contact over the past 21 days. Users must register using their phone number and enable Bluetooth. Phones with the *TraceTogether* app will exchange Bluetooth signals with other nearby phones that have the same app installed. Phones without the app are not detected. The close-contact records are stored on each user's phone, encrypted, and are not uploaded to the network. If a user is diagnosed with COVID-19, the government may request them to provide the data in order to track individuals who have had close contact with them. Additionally, relevant authorities will promptly notify people who have been in close contact with confirmed cases, so they can monitor their health for any symptoms and seek guidance or care as needed. *TraceTogether* does not collect or use any user location data, nor does it access users' contact or address information. It uses Bluetooth data to establish connections without storing information about the location of those connections. Moreover, when two phones exchange information, they communicate using randomly generated IDs, and no personal identity information is involved. The pairing of phone numbers with IDs is stored on a separate server and can only be accessed when necessary. The data is stored on the phone for 21 days and is automatically deleted after that period (Open Source China, 2020).

3 Points of Interest

3.1 Introduction

Points of Interest (POIs) is a common concept in the field of Geographic Information Systems (GIS). It generally refers to geographic entities that can be abstracted as "points," especially those closely related to people's daily lives, such as bus stops, post offices, restaurants, hotels, banks, ATMs, gas stations, hospitals, and supermarkets. POIs describe the spatial location of geographic entities as well as their related attribute information, significantly enhancing people's ability to describe and query geographic features.

A typical POI includes basic information such as name, category, longitude, and latitude. POI data is essential for various electronic maps and has many practical applications, such as assisting users with navigation, providing detailed information about the surrounding area, and helping to plan travel routes.

To facilitate data recording and differentiation, POIs are typically categorized. Each category corresponds to a specific industry code and name. In the August 2017 version of the AMap custom encoding format, POIs are divided into 23 major categories, 264 sub-categories, and 869 finer categories. The 23 major categories include: automobile services, automobile sales, automobile maintenance, motorcycle services, catering services, shopping services, life services, sports and leisure services, healthcare services, accommodation services, scenic spots, business residential areas, government agencies and social organizations, scientific and cultural services, transportation facilities, financial and insurance services, companies and enterprises, road ancillary facilities, place names and address information, public facilities, events and activities, indoor facilities, and access facilities. Table 2 shows some of the POI category codes from AMap, along with major, sub, and finer category names.

Table 2 Examples of Amap POI classification

Type code	Major category	Subcategory	Specific category
050000	Food Services	Chinese Restaurant	Hunan Cuisine (Xiang Cuisine)
070000	Life Services	Telecommunications Service Hall	China Telecom Service Hall
090000	Medical and Healthcare Services	Specialized Hospital	ENT Hospital
130000	Government and Social Organizations	Justice Institutions	Procuratorate
140000	Education and Cultural Services	Media Organizations	Publishing House
150000	Transportation Services	Airport Facilities	Shuttle Bus Station

Source https://developer.amap.com/api/webservice/download

The collection of POI data is a time-consuming and labor-intensive task. Traditional geographic data collection requires professionals to use precise surveying instruments to obtain the latitude and longitude of a POI and then mark it. Currently, there are many specialized electronic map providers worldwide that collect POI data and offer corresponding services, such as AMap and Baidu Maps, which we commonly use. Additionally, many volunteers upload POI data to open-source map websites, collaborating to organize and edit the data online for public use. One of the most well-known examples is OpenStreetMap (OSM). Launched in July 2004, OSM aims to create and provide freely accessible geographic data (including POI data) for anyone who wants to use it, similar to a map version of Wikipedia. Users can directly download global POI data via OSM's data interface to assist with real-world analysis and research. Since the number of OSM users in China is relatively small, the POI data collected may not fully represent the actual spatial information of Chinese cities. Therefore, POI data from AMap or Baidu Maps is often preferred. For example, AMap offers POI information through keyword search, surrounding area search, polygon search, and ID query.

Users can obtain POI data by calling web map APIs or directly accessing open POI data. In addition to OpenStreetMap, some domestic institutions also provide free or paid POI data download services. For instance, the East China Normal University Humanities and Social Sciences Big Data Platform (http://sdsp.ecnu.edu.cn/sdp) offers nationwide Baidu Map POI data covering the period from 2012 to 2017. Users need to request permission to download the dataset. Peking University's Open Research Data Platform (https://opendata.pku.edu.cn) provides AMap POI data, which is offered by a cooperating company, Beijing GuoXin HongShu Technology Co., Ltd., in collaboration with the National Information Center. This dataset covers full data up to September 30, 2018, and includes over 65.3 million entries, with eight core fields: POI name, category, address, latitude and longitude, province name, city name, district name, and district code. After registering and logging in, users can download the data files.

3.2 Research and Application

The Impact of Living Environment Factors on Leisure Activities. POI data can represent urban land-use function types. Liu et al. (2020) systematically analyzed the characteristics of leisure activities in Beijing and the influence of socio-economic and living environment factors on these activities. Their research utilized 1.2 million POI data points collected through the AMap API to characterize the living environment factors in Beijing. The authors selected specific POI types that might influence leisure activities, including parks, sports venues, tourist attractions, and shops. The results showed that Beijing residents prefer to engage in leisure activities within walking distance of their homes, with activity areas primarily concentrated in the city center or district/town centers. This differs from trends in some Western countries, where more than half of activities in certain German cities may occur in places farther from home or difficult to reach by walking. Parks and tourist attractions positively influence leisure activities, which aligns with our understanding of these spaces as naturally suitable for leisure. However, the density of sports venues did not have a significant impact on leisure activities. This may be because the study involved the general population of Beijing, a large proportion of whom prefer to engage in leisure activities in free public spaces such as squares and park open areas (e.g., square dancing or playing chess), which are more accessible than sports venues. Since sports venues usually charge fees, have restrictions on the type of activities, and require certain abilities or physical conditions, they are more popular among younger people and specific groups, rather than the general population, hence showing no significant impact at the city-wide level.

4 Social Media Data

4.1 Introduction

Social media refers to internet platforms based on user relationships for content production and exchange. Common social media platforms include Weibo, WeChat, blogs, Douyin, and international platforms like Facebook, Twitter, Instagram, and Flickr. Users share opinions, experiences, news, and other content on these platforms, which may be presented in various forms, including text, images, audio, and video.

Social media data refers to the data generated by users when they produce and exchange content on these platforms. In actual research, widely used social media data includes user attribute data, user relationship data, geographic location data, and text data. Data related to images, audio, and video is growing, but it is relatively underutilized in urban governance research. Social media data from Weibo, Twitter, and Facebook are the most commonly used in research fields.

User attribute data is the foundational data for research, mainly including information such as user gender, geographic distribution, interests, and the devices used for posting. User relationship data primarily refers to information about mutual followings between users. From an urban research perspective, we often further analyze users' geographic locations. For example, by combining geographic distribution information, we can map the strength of connections between different regions and analyze the characteristics of social networks and urban networks based on regional units. Geographic location data refers to the location information tagged by users when they post on platforms like Weibo, often called "check-in" information. Text data refers to the textual content posted by users, which can be mined to reflect more subjective information such as users' moods and opinions.

4.2 Research and Application

Analysis of Hurricane Sandy Event Using Twitter Big Data. The Data Mining and Machine Learning Lab at Arizona State University developed TweetTracker and TweetXplorer to study Twitter's big data. TweetTracker is responsible for the real-time collection and processing of Twitter data, which can track tweets related to significant events based on hashtags, keywords, geographic boundaries, usernames, and other information, such as news about the U.S. elections (Kumar et al., 2011). As of now, TweetTracker has collected and processed more than 3.78 billion tweets. TweetXplorer primarily handles the analysis, display, and management of Twitter data. It effectively helps researchers understand the "4W" problem: when, where, what, and who. By analyzing tweets from specific important users during a defined period in certain locations, researchers can better understand key information about related events.

Morstatter et al. (2013) explained how TweetXplorer was used to understand Hurricane Sandy. Hurricane Sandy was a Category 1 hurricane that formed over the Atlantic Ocean and made landfall in Cuba, the Dominican Republic, Jamaica, and the U.S. in October 2012, causing significant casualties and property damage. The authors collected a total of 5.64 million tweets related to Hurricane Sandy from October 25 to November 3, 2012. The tweets were grouped into three phases: before the hurricane's landfall in the U.S. (Oct 29, 00:00–Oct 29, 17:59), during the hurricane's landfall in the U.S. (Oct 29, 18:00–Oct 30, 23:59), and the recovery phase after the hurricane (Oct 31, 00:00–Nov 1, 12:00). In the first phase, the most retweeted tweet was about the availability of pet shelters in evacuation zones. As the hurricane approached landfall, users discussed topics related to rumors, damage, and subways. After the hurricane made landfall in the U.S., the first reports were about floods, along with related image links. Subsequently, power outages were reported in New York City and surrounding areas, and some hospitals began evacuating patients. Meanwhile, rumors began to spread, with claims that the flood had submerged the New York Stock Exchange building. Some institutions (like @weatherchannel) retweeted this information but retracted it after confirming the report was false. During the recovery phase, users' main focus shifted to post-disaster repair efforts, with much discussion in New York City about damage, power, and floods.

5 Trajectory Data

5.1 Introduction

A trajectory can be seen as the trace left by a moving object over time in geographical space. Trajectory data has spatiotemporal characteristics and is generated by sampling the movement process of the object. It typically includes information such as the location of the sampling points (longitude and latitude), sampling time, and the object's movement speed. Devices with GPS (Global Positioning System) functionality, such as smartphones, mobile communication base stations, and bus cards, can generate trajectory data. To some extent, the previously mentioned smart card swipe data, mobile big data, and check-in data from social media like Weibo can also be categorized as trajectory data. However, these data have longer sampling intervals and generally broader spatial spans. For example, two consecutive check-in data points for a user might be three days apart, with one from their visit to the Forbidden City in Beijing and the other from their home in Shanghai. This section focuses on more continuous trajectory data, which can more clearly reflect the user's continuous spatiotemporal marks. The most typical example of this data is the GPS trajectory data from taxis. With the widespread use of mobile devices with GPS functionality (mainly smartphones) and the development of location-based services (LBS), a large amount of trajectory data is generated and used for various applications.

Raw GPS trajectory data often contains a lot of redundancy and noise. After a series of data preprocessing steps, the trajectory data can be refined for further analysis. For instance, in the research on identifying bike-sharing stations, Zhang et al. (2019) used the following three main steps to process the trajectory data generated by user rides:

(1) Resampling: The time precision of shared bike trajectory data can reach the level of seconds. However, due to storage and computational limitations, such fine-grained location data is not necessary for data mining, as it can be redundant. Therefore, resampling of the original trajectory data is performed to remove some redundant GPS points while retaining key location information.
(2) Stay Point Filtering: Shared bikes may remain in a specific location for a longer time, such as at temporary parking spots. This location is referred to as a "stay point." Since shared bikes are still in use, a large amount of GPS data can be generated at these stay points, leading to redundancy. This step aims to extract the stay point information and retain only the necessary location data.
(3) Map Matching: This step maps the trajectory sampling points to a real-world road network, estimating the actual user's cycling route. Road network matching plays a significant role in assessing traffic flow, guiding shared bike navigation, and predicting cycling routes.

5.2 *Research and Application*

Identification of Urban Black Holes and Urban Volcanoes. In a research paper, Hong et al. (2015) from Microsoft Research Asia introduced the concepts of urban black holes and urban volcanoes. An urban black hole refers to a situation where the number of people entering a certain area within a given time is much greater than the number exiting, resembling a black hole that can absorb everything, including light (a concept from astronomy). In contrast, an urban volcano is the opposite, where the number of people entering an area is far less than the number leaving, akin to a volcano rapidly releasing a large amount of material (a geological concept). Real-time and accurate identification of urban black holes and volcanoes can assist with urban planning and governance. For example, the serious crowd stampede incident that occurred on December 31, 2014, at the Bund in Shanghai. At the time, many tourists and locals gathered at the Bund to welcome the New Year. The large crowd caused someone to lose balance and fall at the bottom of a stairway on a pedestrian path, which led to a chain reaction of falls and overcrowding, resulting in a stampede. In this incident, the Bund quickly attracted a large crowd, which could be seen as a kind of urban "black hole." If this abnormal situation had been identified earlier, city management authorities could have used information such as crowd density, time, and location to arrange more police officers to maintain order, guide the crowd, limit gatherings in advance, and issue appropriate warnings about potential risks, thereby helping to prevent such public safety incidents. Using a spatiotemporal graph-based method, Hong et al. (2015) applied approximately 18 million GPS trajectory data points from about 33,000 taxis operating in Beijing in November 2012 to identify urban volcanoes and black holes. They have identified urban "black holes" and "volcanoes" in places such as the Workers' Stadium in Beijing, Beijing South Railway Station, Beijing West Railway Station, and Sanlitun. These locations are often near transportation hubs, sports event venues, and entertainment and shopping centers, which makes them prone to attracting large crowds in a short period of time.

Assessment of the Energy and Environmental Benefits of Shared Bicycles. As a sustainable mode of transportation, the widespread adoption of shared bicycles can bring a range of social, environmental, and economic benefits to cities. Zhang and Mi (2018) utilized vast trajectory data generated by user cycling activities to scientifically assess the potential environmental and energy benefits of the new generation of dockless shared bicycles based on internet rental systems. The data used in the study came from approximately 1.02 million anonymized shared bicycle trip records in Shanghai from August 2016. Each record includes the starting and ending locations of the trip, as well as GPS data points during the ride, which can be used to infer the user's true cycling trajectory and estimate the actual trip distance. Zhang et al. then hypothesize a simplified scenario where trips exceeding a certain distance could potentially be replaced by car travel. Since car trips incur corresponding energy and environmental costs, the potential energy and environmental benefits of these longer shared bicycle trips can be represented by the energy and environmental consumption that would result from the replaced car trips. Using data from August as a baseline, the study estimates the total energy and environmental benefits of shared bicycles in Shanghai for the entire year of 2016. The results indicate a reduction of approximately 8,358 tons of gasoline consumption, 25,240 tons of carbon emissions, and 64 tons of nitrogen oxide emissions. From a spatial perspective, the environmental benefits are mainly concentrated in the densely populated central areas of Shanghai, gradually decreasing towards the suburbs. From a temporal perspective, the environmental benefits are most significant during peak commuting hours in the morning and evening. The results of the spatiotemporal analysis are generally consistent with the areas where shared bicycles were deployed and the usage scenarios. This paper is the first to quantitatively assess the environmental and energy benefits of the new generation of dockless shared bicycles, and its findings provide scientific evidence and policy recommendations for the planning and governance of sustainable, low-carbon, and pedestrian-friendly cities.

Identification of Shared Bicycle "Biking Islands." Planning for pedestrian-friendly transportation systems plays an important role in promoting sustainable urban mobility. Traditional urban cycling concepts tend to encompass terms like cycling networks and bicycle parking points, which, from a geometric perspective, can be viewed as "lines" and "points," respectively. However, in some planning for pedestrian-friendly transportation, the concept of "areas" (or "surfaces") is also considered. For example, in the 2006 "Shanghai Central Urban Area Pedestrian-Friendly Transportation System Plan" developed by experts from Tongji University, the proposal was made to create more than 300 urban charm zones (pedestrian hubs) and dozens of urban pedestrian safety zones (cycling islands) scattered across the central urban area of Shanghai. Zhang et al. (2019) introduced the new concept of "biking islands," which refers to areas with higher cycling demand compared to the surrounding regions. If we represent cycling demand in three-dimensional space using height, the demand in these areas would be higher than in the surrounding regions, akin to an island rising above the sea surface, hence the term "biking island."

The data used in the study consisted of shared bicycle trajectory data covering the whole of Shanghai, with a total of 720,000 trip records generated by 130,000 users, involving 290,000 shared bicycles. The data spanned from September 16 to 30, 2017. Each trip record includes a trip ID, start/end times, coordinates of start/end locations, and GPS trajectory sequences between the start and end locations. The basic methodology employed by the authors included resampling, filtering stay points, and map matching, combined with cycling route networks obtained from OpenStreetMaps. The user cycling trajectory data were processed by matching each trip to the bicycle route network, allowing the number of trips on each road segment to be determined. Then, based on the percolation theory in physics, biking islands were dynamically identified. This study contributes to a better understanding of the spatial structure of cycling spaces and supports the planning of urban pedestrian-friendly transportation systems.

References

De Montjoye, C., Radaelli, L., Singh, V. K., et al. (2015, March 18). Unique in the shopping mall: On the re-identifiability of credit card metadata. *Science, 347*(6221), 536–539. https://doi.org/10.1126/science.1256297

Hong, L., Zheng, Y., Yung, D., et al. (2015). Detecting urban blackholes based on human mobility data. In *GIS: Proceedings of the ACM International Symposium on Advances in Geographic Information Systems*. Association for Computing Machinery. https://doi.org/10.1145/2820783.2820811

Jia, J. S., Lu, X., Yuan, Y., et al. (2020, March 18). Population flow drives spatio-temporal distribution of COVID-19 in China. *Nature, 582*(7812), 389–394. https://doi.org/10.1038/s41586-020-2284-y

Kumar, S., Morstatter, F., Sampson, J., et al. (2011, March 19). TweetTracker: An analysis tool for humanitarian and disaster relief. *The International Conference on Weblogs and Social Media.* http://tweettracker.fulton.asu.edu/Kumar-etal_TweetTracker.pdf

Liu, Y. Q., Zhang, Y. P., Jin, S. T., et al. (2020). Spatial pattern of leisure activities among residents in Beijing, China: Exploring the impacts of urban environment. *Sustainable Cities and Society, 52*(July 2019), 101806. https://doi.org/10.1016/j.scs.2019.101806

Morstatter, F., Kumar, S., Liu, H., et al. (2013). Understanding twitter data with TweetXplorer. In *Proceedings of the ACM SIGKDD International Conference on Knowledge Discovery and Data Mining* (pp. 1482–1485). Association for Computing Machinery. https://doi.org/10.1145/2487575.2487703.

NetEase Technology. (2017, February 13). *China Unicom's Smart Footprint: A rapidly growing location-based big data service company.* https://tech.163.com/17/0213/15/CD5S10J900094PRL.html

Newzoo. (2020, March 8). *Global mobile market report (2020).* https://newzoo.com/insights/trend-reports/global-mobile-market-report-2020-light-chinese/

Open Source China. (2020, March 18). *Singapore to release the underlying technology of the open-source COVID-19 app, TraceTogether*. Sina Tech. https://tech.sina.cn/2020-03-28/detail-iimxxsth2224193.d.html

Smart Footprint. (2021, March 18). *Smart Footprint—Urban planning*. http://www.smartsteps.com/index.php

Zhang, Y., Lin, D., & Liu, X. C. (2019). Biking islands in cities: An analysis combining bike trajectory and percolation theory. *Journal of Transport Geography, 80*(2), 102497. https://doi.org/10.1016/j.jtrangeo.2019.102497

Zhang, Y., & Liu, L. (2019). Temporal point pattern analysis of human activities using GIS methods: A case study of library visiting activities in Chengdu city China. *Professional Geographer, 71*(4), 738–750. https://doi.org/10.1080/00330124.2019.1611456

Zhang, Y., & Mi, Z. (2018). Environmental benefits of bike sharing: A big data-based analysis. *Applied Energy, 220*(March), 296–301. https://doi.org/10.1016/j.apenergy.2018.03.101

Zhang, Y., Song, Y., Zhang, W., & Wang, X. (2024). Working and residential segregation of migrants in Longgang City, China: A mobile phone data-based analysis. *Cities, 144*, Article 104625.

Zhang, Y., Zhang, H., Wang, C., Song, Y., & Zhang, W. (2025). Age segregation analysis of three Chinese cities using random walks and mobile phone data. *Cities, 158*, Article 105652.

Zhang, Y., Zhang, Y. P., & Zhou, J. (2021, March 15). A novel excess commuting framework: Considering commuting efficiency and equity simultaneously. *Environment and Planning B: Urban Analytics and City Science, 48*(1), 151–168. https://doi.org/10.1177/2399808319851517

Chapter 4
Case Studies on Technology-Driven Urban Governance

1 City Brain

1.1 Background

In recent years, under the leadership of the innovation-driven development strategy represented by big data, cloud computing, and artificial intelligence, China has been transitioning from the information technology era to the data era. Empowering urban governance through social governance innovation, and boosting the modernization of the national governance system and governance capabilities, has gradually become a hot topic. The core of this approach is to leverage information technology and data processing technology to reform traditional social governance methods and public service modes, thus establishing a social governance system based on co-construction, co-governance, and shared benefits. The goal is to improve government efficiency, enhance the quality of public services, and, ultimately, increase the sense of gain, happiness, and security of the people.

Hangzhou, closely focusing on the strategic goal of modernizing the national governance system and governance capacity, innovatively started the construction of the city brain in 2016, aiming to systematically and scientifically push forward the work of traffic congestion control by utilizing the big data of the traffic department. Today, the Hangzhou City Brain has completed the development of 11 major systems and 48 application scenarios, covering areas such as public transportation, urban management, health and sanitation, and grassroots governance, with an average of 120 million data points being processed daily. In the process of prevention and control of the new pneumonia epidemic, Hangzhou City Brain has provided strong support for the integration of epidemic prevention and control and economic and social development through precise and intelligent control and accurate policy implementation according to the changes of the epidemic. Today, the Hangzhou City Brain has realized three stages of development, from experience accumulation to governance system construction to governance system formation, which corresponds to

W. Zhang and Y. Zhang, *Digital Urban Governance in China*, Qizhen Humanities and Social Sciences Library, https://doi.org/10.1007/978-981-96-6325-5_4

the three stages of digital traffic jam control, digital city control and digital epidemic control.

Practice has proven that as a new urban infrastructure, the Hangzhou City Brain is a key tool for improving social governance capabilities and an essential part of advancing the modernization of the national governance system and governance capabilities. Only by persistently adhering to social governance innovation can the well-being of the people be continually enhanced, urban sustainable development be promoted, a society of freedom and equality be nurtured, and a civilized and harmonious nation be established.

1.1.1 The Genesis of the "Jack Ma's Tardiness Incident"

Since 2010, with the significant improvement of Hangzhou's economic status and the booming development of tourism, urban congestion has drawn attention from all sectors, and tourists have jokingly referred to the city as the "traffic jam city". During holidays, rainy weather, or peak commute hours, traffic jams can stretch for several kilometers, and vehicles may travel only a few kilometers in an hour. On December 26, 2012, at the China Private Enterprise Summit, a prominent figure, Jack Ma, attracted attention for being late. CCTV host Dong Qian asked him: "Why is Jack Ma late?" Jack Ma apologized, saying, "I didn't expect the traffic jam from Yuhang District of Hangzhou to the Great Hall of the People of Zhejiang Province to last for 1 h and 20 min." Yuhang District is located in the suburban area of Hangzhou, and normally, it would take about 30 min to reach the city center.

The then Zhejiang Provincial Party Secretary, Xia Baolong, took this seriously and immediately apologized to Jack Ma, saying, "Traffic congestion is not your problem, but mine." He continued, "When I was governor, I made a decision that within five years, we would tackle the traffic congestion across Zhejiang Province. This time, Vice Governor Wang Jianman has continuously gathered research from all sides, and now we have a preliminary idea." Xia Baolong stated, "I believe that in five years, when Jack Ma travels from Yuhang to the People's Congress Hall, it will take only 20 min." At the same time, Xia Baolong also expressed that Zhejiang needed to address the "traffic" issue in economic development, and that the government would increase efforts for reforms. This preliminary idea marked the beginning of the concept of the "City Brain". In 2016, as part of a 5G-connected vehicle pilot project, the Hangzhou City Brain project was introduced to the Yunqi Town for exploratory development. Meanwhile, Academician Wang Jian, one of the founders of Alibaba Cloud, led this project as the chief architect of the City Brain's technology.

1.1.2 Traffic Congestion Becomes the Primary Problem in Hangzhou

Traffic difficulties and parking issues have long been global challenges hindering the development of large cities. As the central city of the Yangtze River Delta and the core city of the Hangzhou metropolitan area, Hangzhou has been facing an unavoidable dilemma of traffic congestion. Since 2012, Hangzhou's traffic police have focused on improving the city's road traffic capacity, constantly exploring potential, and innovating methods to systematically and scientifically advance traffic congestion alleviation efforts. They have worked to maximize the allocation of police resources while vigorously promoting intelligent traffic construction as an important tool for assisting law enforcement, controlling congestion, and managing disorder. However, due to insufficient utilization of urban resources, the city faces direct or indirect challenges, such as prominent contradictions in road resource supply, lack of coordination in comprehensive governance, underuse of intelligent traffic systems, and difficulty in integrating data resources. These issues prevent existing traffic congestion measures from aligning with the evolving trends of road traffic development in a timely and effective manner. Some innovative measures have even failed to sustain their impact or have been prematurely "terminated" due to dramatic changes in traffic flow patterns and other congestion "factors".

Combined with the initial idea put forward by Secretary Xia Baolong, Hangzhou started to explore the use of large-scale data to improve urban traffic in 2016, and the Hangzhou City Brain Traffic System Version 1.0 was formally released in October 2017, aiming to give full play to the experience advantages of the public security and traffic police departments and the technological advantages of enterprises such as Alibaba, to innovate concepts and tackle problems in science and technology, and to explore paths and methods for the promotion of the modernization of the city's urban governance system and governance capacity. According to the Traffic Analysis Report of China's Major Cities, Hangzhou's ranking in the 100-city congestion index has dropped from No. 3 in 2013 to No. 35 in 2018 (Gaode Maps, 2020). One of the most important reasons for this significant decline is the effective implementation of a series of "smart congestion management" efforts represented by the City Brain application.

1.1.3 Increasingly Prominent City Issues

As the process of urbanization accelerates, while bringing convenience to the people, cities also face an increasing number of challenges, such as traffic congestion, inefficient management, resource waste, environmental pollution, housing difficulties, and challenges in accessing healthcare. The traditional approach of relying on large amounts of human and material resources to govern cities is increasingly difficult to sustain in modern society, and urban transformation faces numerous challenges. In December 2018, the comprehensive version of the City Brain 2.0 was launched, covering nine key areas related to people's livelihoods, including parking, healthcare, culture and tourism, and grassroots governance. The release of the City Brain 2.0 marked the expansion of the City Brain from a single traffic congestion management system to a comprehensive platform that serves public welfare and supports decision-making. This also marked Hangzhou's official shift from focusing solely on "traffic congestion management" to comprehensive "urban governance" through digital urban governance. City Brain 2.0 aimed to integrate data from more departments into the platform, allowing it to unify data processing, analysis, and coordination. It enabled a broader and higher-level view of urban management and made data resources the city's core competitive advantage through computation, and continuously enhanced the ability of managers to govern and address various "urban diseases".

1.1.4 Fragmentation of Urban Governance Capacities Becomes Prominent

Since the implementation of the City Brain Comprehensive Version, data governance has become an essential support for leading economic and social innovation, with data emerging as the core resource for urban development. However, during the exploration of digital urban governance, the issue of fragmented information system construction, with each system operating independently, has become increasingly prominent. In cases of data anomalies, there are delays in early warnings or decision support, and blind spots in urban management remain. The phenomenon of "data islands" remains unresolved. To address this fragmentation and realize centralized urban governance, Hangzhou developed and implemented the Digital Cockpit System. The Digital Cockpit is a service management platform that provides data-driven, online, and intelligent support for urban governance based on the central system's computational power. Once data is interconnected, the Digital Cockpit can present real-time data to urban managers, enabling them to access first-hand information and public sentiment. By comparing and analyzing the data, it provides direction and reference for decision-making, thereby supporting the making of scientific and timely decisions. On September 30, 2019, the Hangzhou City Brain Digital Cockpit was officially launched. The digital cockpits for the city's mayor and party secretary, as well as several systems such as health, culture and tourism, urban management, and finance, were made operational.

1.1.5 COVID-19 Pandemic Control and Resumption of Work and Production

During the COVID-19 pandemic control process, Hangzhou's City Brain precisely controlled the situation and implemented targeted measures based on the evolving nature of the pandemic, providing strong support for coordinating epidemic prevention and economic and social development. In the early stages of resuming work and production, Hangzhou pioneered the "Enterprise Resumption Digital Platform" and the "Hangzhou Health Code". On February 9, the first version of the Hangzhou Health Code was officially released, marking the establishment of a unified digital declaration platform for the entire city. Leveraging Alibaba and WeChat's digital platforms and the "Government Cloud" platform, the Hangzhou Health Code, based on the City Brain, rapidly advanced classification, dynamic, and precise governance

(Shi & Ma, 2020a; Wang, 2020). Meanwhile, the launch of the national integrated government service platform's epidemic health information code marked the nationwide spread of this innovation. As of December 19, a total of 4.5 billion interactions, including citizens displaying codes, non-intrusive code verification, and health services, had been completed through the Hangzhou Health Code based on the City Brain (Tianmu News, 2020). The Hangzhou Health Code based on the City Brain addressed the government's deficiencies in information collection and analysis, saving significant governance costs and greatly improving governance efficiency, especially in the precise management of public risks. During the stage of accelerating urban recovery, Hangzhou also explored the creation of the "Qinqing Online" digital platform. The "Qinqing Online" platform, relying on the central system of the City Brain, enables transformation in five areas—government-business communication, government services, policy formulation, policy implementation, and policy outcomes—through process reengineering, data collaboration, and online interaction, ultimately enhancing the effectiveness of government services.

1.2 Key Approaches

1.2.1 Building a Comprehensive Organizational Structure

The construction of the City Brain is a highly complex system project, and thus establishing an effective organizational structure is a key starting point. The Hangzhou Municipal Party Committee and Municipal Government have placed significant emphasis on the organizational structure of the City Brain. From the beginning of the project, they took a holistic approach, coordinating all relevant departments and elements to ensure the concentrated and efficient use of resources. In terms of the overall framework, Hangzhou's City Brain follows the "five-in-one" strategic layout proposed in the 19th National Congress of the Communist Party of China, which integrates political, economic, social, cultural, and ecological aspects. Based on this, a "531" framework system has been gradually established. "Five" refers to five ones, that is, to build "a unified network", to create "a large cloud", to aggregate "a single data repository", to construct "a central hub", and to strengthen "a brain". "Three" refers to three interconnections, that is to realize connectivity between city, district, and department; between central hubs, systems, and platforms; and between government, enterprises, and markets. "One" refers to a single infrastructure, that is to construct the City Brain as a new type of urban infrastructure. Under the guidance of this overall framework, the City Brain consists of three parts: the central brain hub, industry systems, and district/county platforms. The central hub is further divided into three platforms: the computational resources platform, data resources platform, and algorithm service platform. As of October 30, 2019, the city brain data resource platform had collected 83.7 billion entries, and 11 municipal departmental systems such as transportation, police, politics and law, urban management, housing management, health care, culture and tourism, environmental protection,

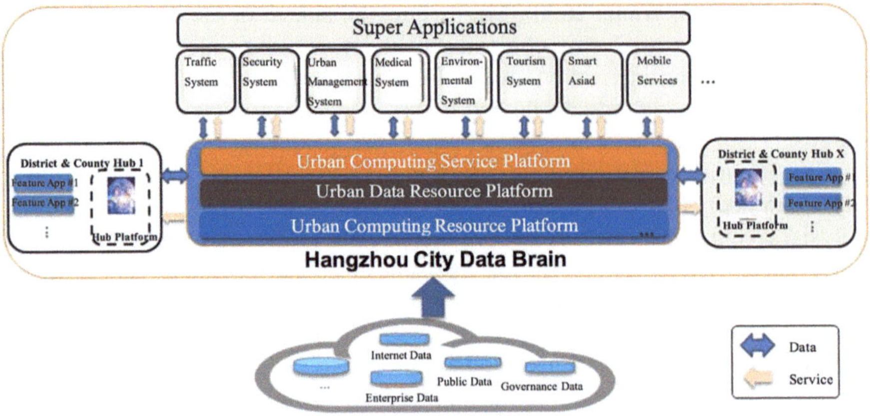

Fig. 1 The framework of the City Brain. *Source* people.com.cn (2019)

agriculture, emergency response, and market supervision, and 17 district and county platforms had been completed, preliminarily forming a general framework with the brain hub platform as the core, and vertically extending to the districts and counties and horizontally extending to the various departments (see Fig. 1).

In terms of specific organizational structure, the Hangzhou Municipal Party Committee and Municipal Government have set up a leadership group and task force for the City Brain. The leadership group is headed by the Municipal Party Secretary, with the main leaders of various departments as members, and an office set up under it, which is responsible for the daily operations by the Data Resource Management Bureau. The task force is divided into city-level departments, county-level task forces, and a comprehensive coordination task force. The divisions are clear, with government officers responsible for streamlining processes, reconstructing workflows, and proposing requirements, while technical teams focus on research, system development, and intelligent implementation. Additionally, the Hangzhou City Brain Operations Command Center was established at Yunqi Town, which integrates operations, command applications, result showcases, and task force R&D. It is managed by the Municipal Data Resource Bureau. A total of 30 task forces are working on the development of the City Brain, including the 11 city-level task forces and 17 district-level task forces mentioned above, along with two key task forces: the Yunqi Engineering Institute and Yunqi City Brain Company. The Yunqi Engineering Institute, composed of volunteers from various enterprises, mainly focuses on developing the central system and data processing iterations. The Yunqi City Brain Company, a state-owned enterprise, is primarily responsible for the daily construction and operation of the City Brain. Driven and pushed by the mechanisms of assessments, regular meetings, information reporting, and work transparency, each task force performs its duties, coordinates closely, and operates efficiently to ensure the successful implementation of the City Brain project.

1.2.2 Creating a Rich Array of Application Scenarios

To comprehensively alleviate urban problems and enhance the sense of access, happiness, and security for the people, Hangzhou's City Brain has created a variety of rich application scenarios, starting with small initiatives that benefit the people. The city has launched numerous scenarios in 11 key areas, including smart traffic, convenient parking, comfortable healthcare, 30-s hotel check-ins, 20-s attraction entrance, digital tourism routes, emergency flood control, vegetable base management, food safety, elevator management, smart housing rental, and smart environmental protection. Many of these scenarios are now in pilot and promotion phases and are gradually being optimized.

At the city-level, in alignment with the concept of "let data run more, and patients run less", the Hangzhou Municipal Health Commission introduced the "One-Time Payment" service to streamline healthcare processes. This service replaces the previous multiple payment steps in clinics and self-service machines with a single payment at the point of care. Following the implementation of "pay after consultation" in 11 city hospitals in April 2019, 245 public medical institutions in the city joined the "City Brain Health System" led by Hangzhou Municipal Commission of Health. The system includes 38 district/county public hospitals, 195 community health service centers, and a provincial hospital (Provincial People's Hospital) (Hangzhou Municipal People's Government, 2019; Zhejiang Online, 2019). Starting from July 1, 2019, patients at these public institutions can choose to pay for their medical expenses either at the hospital after treatment or remotely through mobile payment. This scenario realizes multi-departmental data synergy and is oriented towards optimizing the patient care process, with clear and specific goals and vehicles.

Additionally, the "City Brain Parking System" task force developed its first convenience service, "Pay After Departure", as part of its core "convenient parking" goal. After redesigning the service, "Pay After Departure", car owners can quickly leave the parking lot without needing to do anything, significantly reducing departure time compared to traditional parking payment methods. Meanwhile, in the "Pay After Departure" service, car owners need to bind their vehicles only once to access parking across the city, providing an extraordinary "one parking lot for the entire city" experience. By April 2021, the "Pay After Departure" service had been integrated into over 3,500 parking lots with nearly 750,000 parking spaces, and used by more than 2 million people (Hangzhou CPPCC News, 2021). In addition, parking lots, including 61 state-owned parking lots in the West Lake and Xixi scenic areas as well as those in Yunqi Town, have further implemented "barrier-free" parking. The "Pay After Departure" scenario enables the integration of data from vehicle management departments and payment data from financial departments. This integration optimizes the parking process and simplifies management procedures.

For example, the Hangzhou Municipal Bureau of Culture, Radio, Television, and Tourism has addressed common tourism "pain points" such as traffic congestion, long queues, and waiting times by focusing on the key application scenario of allowing tourists to "explore for an extra hour". To this end, four convenient services have been introduced: "Find a Room in 10 Seconds", "Enter a Scenic Spot in 20 Seconds", "Hotel Check-In in 30 Seconds", and "Digital Tourism Special Line". The "Find a Room in 10 Seconds" service, launched as a WeChat mini-program, addresses the difficulty of finding accommodations during holidays. The front-end platform connects with 8,561 hotels (*People's Daily Online*, 2020a), displaying real-time availability of hotel rooms. Meanwhile, the back-end inventory management system integrates with platforms such as PMS and OTA, enabling tourists to find a suitable room within 10 s based on real-time location and price preferences. The "Enter a Scenic Spot in 20 Seconds" service eliminates the need for tourists to queue at ticket counters. Visitors can use Alipay to scan a payment code at the entrance gates, reducing the average time required for entry from 1.5 min to just 20 s. To date, 22 scenic spots in the city have completed upgrades to their entry gates. The "Hotel Check-In in 30 Seconds" service involves the installation of self-service check-in kiosks in hotel lobbies. These kiosks integrate six systems, including hotel PMS, police registration, access control, payment transactions, OTA reservations, and direct hotel sales. Following this upgrade, the time required for self-service check-in has been reduced to just 30 s, compared to the previous 270 s, saving approximately 4 min. Currently, 75 hotels in the city have completed self-service upgrades (Zhejiang Online, 2019). The "Digital Tourism Special Line" service, in collaboration with enterprises such as the Public Transport Group and Microbus, aggregates various data types, including public transportation operations, railway reservations, and tourist trajectories, into the City Brain system. Through correlation analysis and computation, it scientifically plans specialized routes and dynamically schedules departure times. This service explores a new model for tourism transportation that matches data with resources and integrates market-oriented operations. As of now, 22 digital tourism special lines have been launched across the city.

At the district and county platform level, the "One-Click Escort" scenario in Xiaoshan District leverages artificial intelligence to enhance emergency response efficiency, transforming legal empowerment into technological enablement. Previously, ambulances with the legal right to run red lights (120 emergency vehicles) were prone to secondary accidents. Now, through collaboration between the City Brain backend and traffic police, traffic lights along the route automatically turn green based on the vehicle's speed and location. This system achieves an average signal control rate of 90%, increases speed by 50%, and reduces travel time by 30% (*People's Daily Online*, 2019), creating a green channel for safeguarding people's lives. Similarly, the construction of the City Brain East Railway Station Demonstration Zone in the former Jianggan District (now part of Qiantang District and Shangcheng District as of March 11, 2021) has launched applications in five scenarios: digital tourism special lines, parking resource optimization, traffic organization optimization, garbage metering and collection, and smart elevators. Among these, the "digital tourism special line" saves tourists traveling to the West Lake Scenic Area an average of 40 min. Parking resource optimization has increased the underground parking lot space utilization index from 2.5 to 3.3. The smart elevator system has saved the government over 119,000 yuan in investments (Zhejiang News, 2019).

1.2.3 Promoting Collaboration Among Government Departments

Through practical exploration, Hangzhou's City Brain has gradually established and continues to refine a social governance system characterized by "Party leadership, government guidance, social collaboration, public participation, and legal guarantees". A leadership group for the construction of the City Brain, headed by the Municipal Party Secretary and composed of leaders from various district and county departments, focuses on fostering departmental collaboration. The newly established Hangzhou Data Resource Management Bureau, serving as the lead organization, coordinates the division of labor among 30 task forces. Disagreements and conflicts of interest inevitably arise during departmental collaboration. If departments adhere rigidly to their respective positions without compromise, progress becomes difficult. In fact, innovating mechanisms and systems to foster genuine cooperation among departments for the construction of the City Brain poses a greater challenge than technological breakthroughs. This effort also tests the modernization of governance systems and capabilities. According to incomplete statistics, Hangzhou previously had 52 government departments and units managing a total of 760 information system projects, creating numerous data silos. Data sharing between departments was nonexistent, and even within the same department, different business lines were often unconnected. The Hangzhou City Brain has broken through these traditional data-sharing and aggregation barriers, constructing a city-scale super data collaboration center (Super Data Collaboration Center). This center provides a centralized solution for collaborative governance across government departments. By offering data and computing services through the central system, it significantly enhances the city's data and system collaboration capabilities while avoiding the issues of centralization or platformization in traditional smart city construction models.

In addition to innovations in the social governance system, Hangzhou's City Brain has achieved new breakthroughs in data integration by leveraging the "Digital Cockpit". In traditional government operations, issues such as information distortion caused by multiple layers of reporting were common. The introduction of the Digital Cockpit effectively addresses these challenges. The Digital Cockpit is a data-driven, online, and intelligent urban governance platform supported by the computing power of the City Brain's central system. By integrating, merging, and computing various types of urban operational data, it enables horizontal connectivity, vertical comparisons, real-time monitoring, and intelligent early warnings for various urban operation indicators. This provides city managers with real-time situational awareness, informed decision-making capabilities, and enhanced command and control. The Digital Cockpit can be accessed seamlessly from city-level platforms to district and county platforms, departmental systems, and grassroots platforms at town (sub-district) levels. Through a central architecture defined by the "one framework, two connections, three collaborations, and direct access" model, the system meticulously calculates digital indicators and establishes layer-by-layer connections. The Digital Cockpit strictly distinguishes user access rights, providing customized data content and interfaces based on the roles and levels of city managers. Currently, 42 Digital Cockpits have been activated across city-level platforms, departmental systems, district and county platforms, and sub-district platforms, with each cockpit available in three versions, large-screen, tablet (Pad), and mobile, to fully support mobile operations. Hangzhou has taken the lead in making "governing a city with a single smartphone" a reality.

1.2.4 Connecting Government and Market for Diverse Governance

Under the traditional governance model, which relied on the government as the sole actor, communication between the government and citizens was one-way and top-down. Empowered by Hangzhou's City Brain and leveraging next-generation information technologies such as the central processing system from Yunqi Engineering Institute, cloud computing, and social network services, enterprises and citizens have now become active participants in social governance. This transformation represents a city-wide innovation movement, enabling cooperation and two-way interaction between the government and the market, thereby generating synergistic effects. Furthermore, governance has shifted from reactive problem-solving to proactive command and guidance.

Through early exploration and practice, the construction of the City Brain has established a government-enterprise cooperation model in which the government provides scenarios and requirements, while enterprise teams conduct joint research, resulting in a series of experiences and achievements. In recent years, Hangzhou has vigorously implemented its "Digital Economy No. 1 Project", leading to the emergence of outstanding enterprises such as Alibaba Cloud, Hikvision, Dahua, and SUPCON. These companies possess an international perspective and globally leading technologies, actively participating in the construction of the City Brain by providing essential talent support and technical guarantees. For example, the smart elevator system of the Hangzhou Market Supervision Bureau and the convenient parking system of the Urban Management Committee demonstrate this cooperation. In the former case, through government-enterprise collaboration, social elevators in commercial complexes and office buildings have been incorporated into the government's real-time supervision system. The market supervision department integrates monitoring and inspection data to intelligently oversee elevator maintenance and inspection statuses in real-time, ensuring effective and controllable elevator operations. By analyzing and applying fault data, it enables early warnings and precise supervision. If citizens become trapped in an elevator, intelligent services such as automatic task assignment, nearby dispatching, and real-time rescue trajectory display ensure their safety during emergencies. These measures prioritize public safety and address immediate citizen needs. They have transitioned from merely identifying and solving problems to integrating enterprise data into public governance, encouraging public participation in social management and enhancing citizens' sense of gain, happiness, safety, and ownership of the city.

To address the issues of complex parking payments and difficulty in exiting parking lots, the City Brain parking system aims to create a city-level parking service platform. It was the first in the nation to introduce the "Pay after Departure" convenient parking service scenario. Developed by Hangzhou City Cloud Company under the commission of the Urban Management Committee, this service is a product of government-enterprise collaboration. Through big data analysis and computation, Hangzhou's City Brain has effectively integrated private parking lot information, enabling the efficient allocation of parking resources citywide and providing convenient parking services for vehicle owners. This process involves property management companies overseeing parking operations and technical equipment companies supporting the technology, all participating in the practical governance of social parking. This improves the allocation efficiency of parking resources, highlights the City Brain's role as a "central hub" in urban intelligent governance, and offers a viable path for broader participation by diverse stakeholders.

In addition, the City Brain provided robust support for efficient communication and collaboration among the government, market, and public during the COVID-19 pandemic. At the onset of the outbreak, leveraging the foundational infrastructure of the City Brain, Hangzhou promptly established a joint health and public security epidemic prevention system and an epidemic cockpit. These systems accurately recorded and publicly released information such as confirmed cases, suspected cases, and outbreak locations. By ensuring the timeliness of epidemic information through interdepartmental collaboration and technological tools, the initial social panic was effectively alleviated. During the early stages of the resumption of work and production, Hangzhou pioneered the "Enterprise Resumption Digital Platform", serving businesses promptly and facilitating the online registration of 240,000 enterprises and over 3 million employees citywide. It also swiftly developed and launched the "Hangzhou Health Code", which enabled "digital health certification, precise personnel management, convenient citywide travel, and efficient enterprise work resumption". As the city accelerated its recovery, the "Qinqing Online Online" digital platform was introduced. This platform allowed businesses to submit demands directly, obtain online approvals, receive policy benefits, access services, and evaluate performance in real-time. From its launch on March 2, 2020, to December 19, the platform implemented 303 business-friendly policies, disbursed 7.15 billion yuan in funding, and benefited 2.69 million enterprises and 799,000 employees (Tianmu News, 2020). During the full recovery of economic and social order, innovative cloud services such as cloud-based investment promotion, recruitment, and contract signing were launched, attracting over 100,000 talents for online interviews. Thanks to the modernized operational foundation laid during the construction of the City Brain, coupled with precise and reliable data, Hangzhou was able to rapidly apply data to pandemic prevention and public services. This not only supported leadership decision-making and efficient allocation of public resources but also ultimately improved the effectiveness of social governance.

1.3 Achievements

In just a few years, Hangzhou's City Brain has progressed from nonexistence to existence, from intensive construction to widespread application, leading the way both nationally and globally. The City Brain has achieved remarkable results during its application process, primarily reflected in the following four aspects.

1.3.1 Effectively Enhancing People's Well-Being

The construction of the City Brain firmly focuses on services that benefit and facilitate the public. Starting with practical issues related to people's livelihoods, it continuously expands the application of scenarios that provide convenience and benefits to citizens, ensuring a tangible sense of gain. The City Brain serves as an effective tool to implement a people-centered development philosophy and improve the quality of life. By integrating various data resources and leveraging intelligent algorithms and self-learning capabilities, the City Brain has promoted the digital transformation of infrastructure, livelihood services, and government services. It provides the public with more inclusive, convenient, fast, and precise public services. The City Brain further utilizes data to generate "chemical changes" through comprehensive integration and computation, offering more precise services to businesses and citizens. This transforms the traditional model of citizens seeking government services into a proactive model where the government delivers services to citizens, effectively enhancing people's well-being.

1.3.2 Comprehensively Alleviating Urban Ills

The City Brain implements General Secretary Xi Jinping's directive to build new-era smart cities. By integrating data resources within the government and between the government and society, and leveraging artificial intelligence and big data technologies, it supports optimizing public resource allocation, macro decision-making and command, event prediction and early warning, and addressing "urban ills". Currently, Hangzhou's City Brain has developed a comprehensive management platform covering all aspects of urban management and has established a series of smart application scenarios, including smart transportation, smart urban management, smart tourism, smart governance, and smart healthcare. These efforts have achieved significant results in alleviating urban issues, enhancing Hangzhou's refined urban management and service levels, and improving citizens' quality of life. Furthermore, refined urban management has improved the efficiency of public resource utilization, effectively reducing congestion in urban centers and the costs of environmental management. For instance, using the City Brain, the average travel time on Hangzhou's 22-km Shangtang Elevated Road has been reduced by 4.6 min, approximately 10%. In Xiaoshan District, 104 intersections with automatically adjusted traffic signals have increased vehicle speeds by 15%, saving an average of 3 min per trip. As of 2020, Hangzhou's permanent population reached 11.936 million, with 3.119 million registered vehicles. Despite 101 km of urban expressways and 210 km of metro lines under construction, and a reduction of nearly 20% in available road space, the City Brain has helped increase average road traffic speeds by 15% (Communist Party of China News, 2020).

1.3.3 Enhancing the Level of Urban Civilization

The City Brain, embodying the concept of "making cities think and life better", not only enhances urban intelligence but also fosters a new urban civilization. Years ago, Hangzhou legislated the requirement for motor vehicles to yield to pedestrians, establishing a positive traffic order. This is just one example among many social orders, all of which lay the groundwork for promoting and implementing the Hangzhou City Brain, creating a social demonstration effect that has been widely accepted and followed by the public. With a foundation of order and a sense of social contract, the Hangzhou City Brain, driven by big data, has further improved the level and system of social governance, thereby elevating the overall level of urban civilization. Hangzhou has become the first city in China to implement "barrier-free parking lots", the first to achieve "ambulances not needing to run red lights", the first to use data calculations to orderly relax "traffic restriction measures", the first to adopt a "delay index" as a key performance indicator (KPI) for traffic police, the first to create the "Digital Park Card", the first to realize "queue-free park entry and hotel check-in", and the first to implement "one-time payment after medical services".

1.3.4 Supporting Hangzhou's Digital Economy

By 2020, the City Brain had integrated with 116 departments, 17 districts, counties, and towns in Hangzhou, and 317 informatization system projects (Communist Party of China News, 2020). Real-time data from 20 public service operators, totaling 77.1 billion entries, were aggregated into the City Brain's Data Collaboration Center. The central system incorporated 4,500 API data interfaces, processing an average of over 120 million data interactions daily (*China Youth Daily*, 2019). This effectively solved the problems of data silos and information fragmentation, optimizing resource allocation and utilization efficiency. As an innovative integration of cloud computing, big data, and artificial intelligence, the City Brain serves as a comprehensive platform that deeply connects and supports the collaborative development of Hangzhou's digital economy, digital society, and digital government. It accelerates the integration of digital industrialization, industrial digitization, and urban digitization, opening new channels for digital economic development and creating new growth points. The City Brain also provides a platform for industrial upgrading and innovation. As Hangzhou's City Brain continues to iterate and improve, the data resources it aggregates will continuously give rise to valuable industry applications, guiding startups to innovate using these resources and accelerating the transformation of innovative achievements.

1.4 Lessons Learned

General Secretary Xi Jinping emphasized in his report at the 19th National Congress of the Communist Party of China the need to strengthen and innovate social governance, establish a pattern of co-construction, co-governance, and shared benefits, continuously meet the people's growing needs for a better life, and constantly promote social fairness and justice. Hangzhou's City Brain represents the best innovative practice of social governance at the urban level and is worthy of in-depth study and close attention.

The City Brain, supported by resource integration and information sharing, drives urban governance from offline to an integrated online-offline model. It shifts from single-department oversight to enhanced interdepartmental collaborative governance, providing robust technical support for precise government decision-making and efficient governance. It has effectively improved urban governance capabilities and levels, earning widespread acclaim from urban residents and various sectors of society. The initiative has garnered media coverage and reports from outlets such as People's Daily, Xinhua News Agency, China National Radio, China Central Television, and China Radio International. It was also awarded the "2018 Annual China Top Ten Social Governance Innovation Award" by Xiaokang magazine of Qiushi.

As a new development and a form of infrastructure, Hangzhou's City Brain is still in its early stages. It is essential to recognize its shortcomings and persistently advance its development. For cities currently building their own City Brain systems, Hangzhou's experience in solving practical problems during its pilot phase offers replicable and scalable practices.

1.4.1 People-Centered Approach as the Mission of the City Brain

The core design philosophy of the City Brain is to center on the people. Through data empowerment, it provides a sense of gain for the public, offers convenience to citizens, and benefits enterprises, making it beloved by the public and relatable to businesses. By adhering to the principle that data resources are "sourced from the people and served for the people", the initiative emphasizes problem- and demand-oriented approaches, proactively responding to the needs of citizens and market entities. It carefully designs various application scenarios that touch the pulse of the city, feel its warmth, and enjoy its services. These efforts effectively address urban governance "pain points" and "blockages" as well as "hotspots" and "difficulties" strongly voiced by the public. Through "small-scale breakthroughs", it drives "significant changes", enhancing urban governance capabilities and public service levels. This has increased the public's sense of gain, happiness, and safety, making the initiative a visible, tangible, and beneficial project for the people—a true project of the heart.

1.4.2 Data and Algorithms as the Fundamental Conditions for the City Brain

Big data is the cornerstone of the City Brain, while computational power and algorithms are its key elements. Urban digital governance must be built on a government public data platform, vigorously promoting two-way integration and openness of diverse data. Through scientific algorithm models and robust computational processing, data fusion and innovation can be effectively achieved, enabling comprehensive perception, global analysis, and intelligent management of urban operations. As the artificial intelligence hub and open operation platform for urban governance systems, the City Brain adopts a "trinity" architectural system comprising the central brain, departmental systems, and district and county platforms. This enables cross-regional, cross-level, and cross-sector data aggregation and interconnectivity. It supports real-time connection, scheduling, and processing of diverse heterogeneous data citywide, provides algorithm services with mainstream deep learning frameworks, algorithm components, and integrated algorithm management, and intelligently generates core indicators and key data for urban operational systems. This ensures the real-time online functionality and overall coordination of digital cockpits across different sectors and levels.

1.4.3 Self-Reform as the Intrinsic Driver of the City Brain

The City Brain has driven the government's transformation from "power-based" governance to "data-based" governance, from "experience-based" decision-making to "intelligence-based" decision-making, and from "static" management to "dynamic" governance. Fundamentally, it represents continuous self-reform aimed at dismantling entrenched departmental barriers and monopolistic interests. Big data enables optimization of administrative processes through information and data sharing, thereby enhancing the efficiency of administrative services. For example, the "Convenient Parking" reform is a governance practice based on breaking down information silos, effectively dismantling privileged class interests and optimizing public service processes, embodying the essence of information sharing. Additionally, the government's self-revolution and shared informatization efforts enhance public trust in the government, boost its credibility, and build confidence while laying a solid foundation for constructing new smart cities.

1.4.4 Government-Enterprise Collaboration as the External Driving Force of the City Brain

The City Brain is an extraordinarily complex system project, with its success hinging on the synergistic efforts of an active government and an effective market. During its implementation, the Hangzhou Municipal Party Committee and Municipal Government played a leading role in planning, coordination, policy support, and application demonstrations. At the same time, they broke away from a solely government-funded investment model by establishing Yunqi City Brain Technology (Hangzhou) Co., Ltd. and creating the 2-billion-yuan "Hangzhou City Brain Industry Fund". By introducing market-oriented mechanisms, they explored and formed a new model of smart cities characterized by government-provided scenarios, enterprise-led collaborative innovation, and optimized resource allocation. This effort strives to build a new social governance system of "co-construction, co-governance, and shared benefits".

2 Health Code

2.1 Background

Since December 2019, several cases of unknown pneumonia with exposure history to the Huanan Seafood Market were identified in hospitals in Wuhan, Hubei Province. These were confirmed as acute respiratory infectious diseases caused by the novel coronavirus (COVID-19). Although COVID-19 was identified in December 2019, it did not receive significant attention until January 20, 2020, when Academician Zhong Nanshan confirmed human-to-human transmission. On February 28, 2020, the World Health Organization (WHO) raised the global transmission risk and impact risk level of COVID-19 from "high" to "very high". Subsequently, the global pandemic situation further developed. According to the daily situation reports published by WHO, as of March 10, 2020, the virus had spread to 110 countries and regions worldwide. In this context, WHO Director-General Tedros Adhanom Ghebreyesus announced on March 11, considering the rapid development of the pandemic and inadequate responses in some countries and regions, that the outbreak constituted a "global pandemic" (*People's Daily Online*, 2020b).

By mid-February 2020, as China's COVID-19 prevention and control efforts progressed, promoting work and production resumption had become an urgent priority. Under conditions of large-scale cross-regional labor migration, rapidly and accurately obtaining critical information such as the health status, travel trajectories, and contact histories of massive numbers of returning workers became the key to orderly and categorized work resumption. In this context, digital epidemic prevention measures, represented by the Health Code, emerged and played a significant role.

At the time, like other regions in China, Yuhang District implemented epidemic prevention management by controlling personnel at checkpoints and using paper

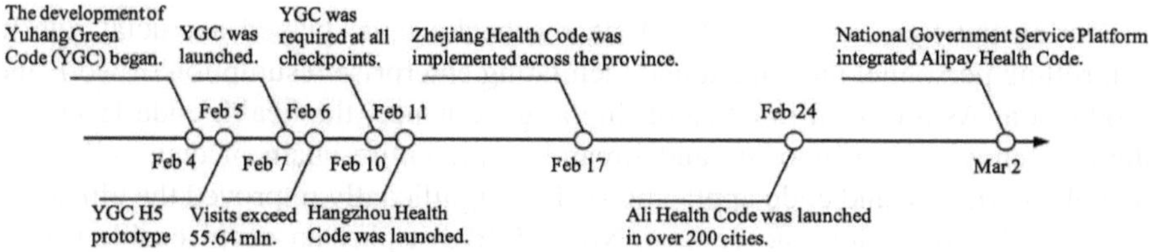

Fig. 2 The timeline of the development of Zhejiang Health Code (Shi & Ma, 2020a)

forms for data collection, which were then aggregated step by step. The pressure on grassroots workers to fill out forms and the demand for a more convenient solution were conveyed to district-level authorities. At a deployment meeting on the afternoon of February 4, the Yuhang District Government explicitly proposed building a digital solution, emphasizing "full population coverage, full-process handling, and full-domain joint prevention". Figure 2 shows the timeline of the development of the Hangzhou Health Code.

On the evening of February 4, Yuhang District formed a 10-member collaborative development team, including technical experts from Alibaba and a small tech company, Maquan Technology. After 12 h, at 5 a.m. on February 5, Maquan Technology's solution was selected as the winner. This became the prototype of the "Yuhang Green Code"—a mobile H5 webpage with basic review functions. Through self-reporting of personal and travel information, it replaced traditional paper certificates and on-site information registration. Additionally, the backend database enabled verification of authenticity and determined whether individuals had contact with epidemic zones or high-risk individuals. From this point forward, the "Yuhang Green Code" became a genuine governance innovation and was submitted to the Hangzhou Municipal Government and the Zhejiang Provincial Government for approval (Shi & Ma, 2020a).

On February 11, Zhejiang Province implemented a travel system using a "red, yellow, green" QR code as a digital health certificate. Residents and visitors to Hangzhou could obtain a health code via market applications such as DingTalk or Alipay, which allowed access to highways, residential communities, and other public places. On February 16, under the policy guidance of the General Office of the State Council, leading internet companies such as Tencent and Alibaba began participating in the construction of a unified national platform for pandemic control and prevention through health information codes. This initiative supported precise epidemic prevention and control and categorized, orderly resumption of work and production across regions. On February 25, the Joint Prevention and Control Mechanism of the State Council issued a notice on scientifically and precisely managing COVID-19 epidemic prevention, encouraging regions with the capability to promote personal health codes and similar information platforms. Among the first 18 provinces to resume work nationwide, 80% had launched health codes. On March 2, the national unified health information code was launched on Alipay Mini Programs, marking the nationwide implementation of this innovation (Zhejiang Daily, 2020).

During the fight against COVID-19, the health code played a crucial role in controlling personnel movement and facilitating enterprise resumption of work and production. As a typical practice of digital governance, the health code leveraged digital technology to innovate and streamline the entire chain of data collection, model algorithms, and code applications. This significantly improved the efficiency of epidemic prevention and control. Not only did it play an irreplaceable role in combating the pandemic, but it also became a vital practice of digital social governance in China, showcasing the country's digital governance capabilities to the world.

2.2 Main Approaches

2.2.1 Technology Empowerment for Dynamic Governance

The health code emerged from the need for information to prevent risks, and technology-enabled dynamic governance is one of its most essential practices and features. Technology empowerment involves not only technical solutions but also an institutional implication of optimizing management efficiency through technical solutions. It also reflects the proactive innovation of local governments, effectively driving the process of meeting emergency needs and innovating government services dynamically. The health code is based on "personal self-reporting, database comparison, and spatiotemporal screening". Using big data technology, it performs "initial instant computation, daily scheduled computation, and dynamic real-time updates" to automatically generate code colors that reflect an individual's health status. Relying on platforms such as WeChat and Alipay, the health code implements a "green code—yellow code—red code" three-color dynamic management system. While its external representation is a colored QR code, its functionality extends far beyond that. The QR code is merely the system's front-end and an entry point for information. The full health code application comprises a platform system, including the front-end interface, cloud computing, interactive interface, algorithmic decision-making backend, and business scenarios (see Fig. 3).

The development of the health code also embodies the concept of dynamic governance. In its initial phase, the health code faced various inquiries, "12345" hotline requests, and complaints. Subsequently, technical teams from different product lines of Alipay, Alibaba Cloud, DingTalk, and others assisted the government in troubleshooting problems, integrating data, and adjusting algorithms, which rapidly improved the accuracy of the determinations. Mature technologies from the e-commerce sector, such as those used to handle massive traffic during "Double 11" sales events and to push targeted advertisements to users, were quickly adapted to the health code application, enabling its integration into governance scenarios. These standardized and packaged technical capabilities are referred to as "middleware capabilities". By empowering previously established e-government databases (the "backend") with "middleware capabilities", a more flexible "government middleware" was

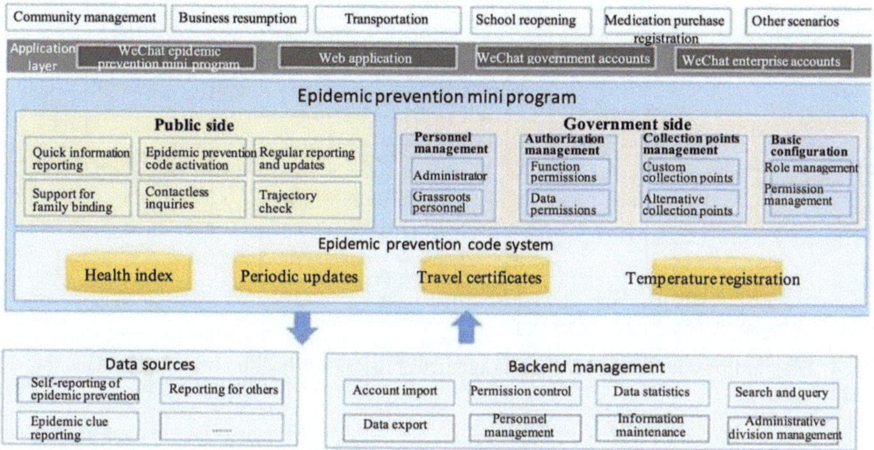

Fig. 3 The technological framework of the health code. *Source* The Paper (2020)

built to better meet "front-end" business needs (Shi & Ma, 2020b). Throughout this process, government applications and scenarios were continuously optimized and upgraded, achieving dynamic governance.

2.2.2 Collaborative Governance by Diverse Stakeholders

The introduction and implementation of the health code were the result of collaborative efforts by the government, market entities, social organizations, and the public. During its rollout, government departments served as the primary driving force. Hangzhou established a dedicated task force for the development and operation of the health code platform. This task force was led and coordinated by the General Office of the Municipal Party Committee, with key personnel from municipal departments such as the Health Commission, Political and Legal Affairs Commission, Development and Reform Commission, Data Resources Bureau, and Economic and Information Technology Bureau participating. The task force was divided into four groups: Comprehensive Coordination, Data Quality, Issue Review, and Planning and Development, which worked together to advance the Hangzhou health code initiative. The government recognized the role of market entities in collecting and analyzing public data, such as location, activities, production, consumption, entertainment, and social relationships, within a regulated framework for epidemic prevention and risk assessment. This was critical for the rapid deployment and use of the health code.

During the development of the health code, market entities leveraged their technological advantages. Platform companies, industry associations, and social organizations actively explored new technologies and methods for data development, conducting in-depth analysis of both structured and unstructured data. This enhanced the value of data utilization and improved the scientific precision and accuracy of risk

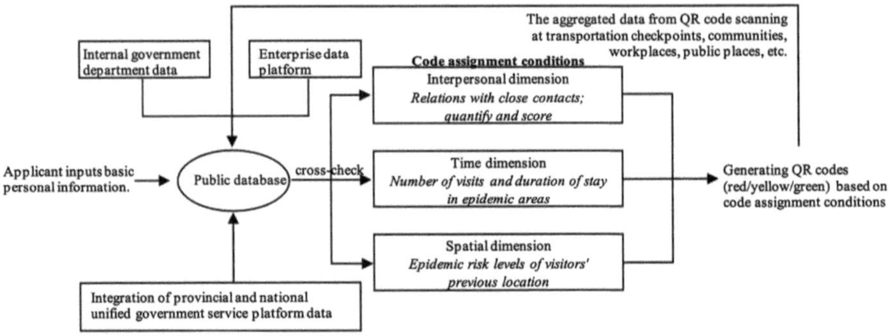

Fig. 4 Diagram of data flow and code assignment process for health code. *Source* Gao and Xu (2020

identification and early warning systems. For example, a proposal by Maquan Technology, a small tech firm in Yuhang District, served as the prototype for the "Yuhang Green Code". Additionally, Alibaba's technical support played a key role in the official launch of the Hangzhou health code. By utilizing data accumulated on the Alipay platform, manual input of information such as names, ID numbers, and phone numbers was no longer required. Alibaba Cloud provided cloud computing support for cross-database comparison of government data, integrating databases related to mobile roaming, health, transportation, and population, enabling automated code assignment after verification (see Fig. 4).

It is also noteworthy that after the health code was implemented, communities played a significant role in its promotion and enforcement. Moreover, the health code's success depended heavily on public support and participation. Citizens supported the platform's collection and use of personal information, particularly their digital data in cyberspace, and accurately participated in the self-reporting of personal information. This public participation was essential to the successful implementation of the health code. Without it, the health code would not have achieved its intended impact. The promotion, deployment, and implementation of the health code represent a process in which government power, market forces, social entities, and individual contributions leveraged their respective strengths to participate in the governance of public risks (Wang, 2020).

2.2.3 Cross-Regional Integration for Mobility Governance

In its early stages, the health code was limited to intra-provincial use and could not achieve cross-provincial mutual recognition. Due to the relatively independent health code systems of different provinces, migrant workers were required to apply for multiple health codes in different locations. The lack of standardized health code generation criteria and the inability to interconnect data between health codes on WeChat and Alipay platforms resulted in issues such as overlapping requirements ("codes upon codes") in some areas, hindering the effectiveness of this emergency innovation. To address this problem, efforts to promote cross-regional mutual recognition of health codes were initiated. First, mutual recognition was established as quickly as possible between areas of population outflow and inflow. For example, Zhejiang and Henan, with a large number of Henan workers employed in Zhejiang, were among the first to sign a health code mutual recognition agreement to facilitate employee returns. Second, mutual recognition within urban clusters was prioritized. In March 2020, the Yangtze River Delta Regional Cooperation Office proposed mutual recognition and data sharing among Shanghai's "Suishen Code", Jiangsu's "Sukang Code", Zhejiang's "Health Code", and Anhui's "Ankang Code". On December 10, 2020, the National Health Commission, National Healthcare Security Administration, and National Administration of Traditional Chinese Medicine jointly issued a notice titled Guidelines for Deepening the "Internet + Healthcare" and the "Five-One" Service Action. This notice explicitly required the implementation of unified health code policies and standards, nationwide mutual recognition, and seamless travel with one code (CNR, 2021). On March 23, 2021, the National Health Commission announced at a press conference that nationwide "one-code access" had been essentially achieved, removing barriers to cross-regional mobility governance.

Mobility governance transcends geographical constraints by relying on information technology to shift offline governance to online platforms, organizing stakeholders via digital networks. Unlike traditional governance, mobility governance ensures the rights of mobile populations, emphasizing shared participation between locals and outsiders. It also highlights the government's role in integrating powers and resources within and beyond regions, focusing on urban networks, intercity connections, and collaboration among different organizations. Under mobility governance, the health code does not target all people in a specific region, effectively avoiding stigmatization of populations in certain areas. However, in practice, there remain deficiencies in safeguarding applicants' rights. Its color-coding and associated behavior restrictions can inadvertently result in discrimination. To address this, many regions began exploring the integration of the health code into routine social governance during phases of eased pandemic conditions. By reducing its monitoring and management functions and enhancing its role as a public service tool, the health code aims to become an instrument of technological empowerment for the public.

2.2.4 Extension to Routine Governance

As the domestic pandemic eased, the health code began transitioning from an emergency response strategy to routine governance, with its functions gradually shifting from epidemic prevention to public services. Due to its widespread usage and integration with social and payment platforms, some cities proposed incorporating the health code into urban public health and medical systems. Functions such as mask purchasing, epidemic clue reporting, fever clinic navigation, hospital bed availability inquiries, and free online consultations were integrated into the health code, providing convenience for residents. In addition, some cities applied the health code to the transportation sector, integrating it with transit codes. The health code also rapidly iterated its technology, launching international versions, student return-to-school codes, enterprise resumption codes, and resident consumption vouchers. These services provided innovative practices for epidemic prevention and management for foreigners in China, school reopenings, work and production resumption, and boosting consumption during the later stages of the pandemic (Shan, 2021).

On February 21, 2020, the Hangzhou health code achieved interoperability with electronic health cards and social security cards. Patients no longer needed to carry physical medical cards or paper medical records. With a single QR code, they could complete procedures such as registration, medical tests, medication pickup, and electronic invoice issuance. On May 15, Hangzhou held a conference to promote the long-term operation of the health code, planning services such as online contract signing, online appointments, and internet-based diagnostic report inquiries as part of the "Internet + Healthcare" initiative (Shi & Ma, 2020b). Guangzhou's "Suikang Code" has been used as electronic identification, Shanghai positioned its "Suishen Code" as a personal service code for citizens' work and daily life, and Zhejiang launched pilot programs for personal credit codes and enterprise codes. In transitioning from crisis management to routine governance, the health code evolved into version 2.0. Beyond being a technical tool for population registration and work resumption, it developed into a daily public service platform, driving a new wave of urban governance innovation.

2.3 *Achievements*

2.3.1 Assisting in the Restoration of Social Order

From a macro perspective, the health code addressed the issue of safe and orderly cross-regional movement amid localized outbreaks and the resumption of work and production. It reduced the possibility of epidemic spread, alleviated labor market supply-demand pressures, and enabled local governments to scientifically assess the overall health status of regional populations and the spread of the epidemic. From a micro perspective, the health code facilitated individual travel, supported enterprises in systematically deploying healthy employees for work resumption, and assisted students in returning to school, thereby promoting the establishment of an economic and social order aligned with epidemic prevention and control.

Firstly, it reduced "form-filling epidemic prevention" and facilitated personal travel. During the initial outbreak of COVID-19, various levels and units often required repeated and timely form submissions. The health code enabled one-time online reporting of health information, which could be used multiple times across various locations, greatly reducing inconvenience for individuals and easing the burden on grassroots staff. The electronic health code replaced paper-based entry permits, making it environmentally friendly and convenient. It simplified checkpoint procedures and enabled "one code for the entire city" or even "one code for the entire country". Individuals could access residential areas, workplaces, shopping malls, hotels, restaurants, train stations, and other locations without repeatedly filling out forms or providing proof. This saved time, reduced the risk of gatherings and infections, and effectively resolved the challenges of entering and exiting places during work, production, and school resumption.

Secondly, it improved the efficiency of COVID-19 joint prevention and control. The introduction of the health code facilitated decision-making, management, and innovation based on data, becoming a key tool for enhancing the modernization of national governance capacity and systems. By enabling individuals to report health information via scanning, relevant government departments could perform real-time monitoring and management, improving statistical efficiency. Standardizing the health code helped address issues such as varying form scopes, inconsistent data, and non-interoperable systems, thereby reducing significant human and material costs. Fully utilizing big data technology through the health code significantly improved the efficiency of joint prevention and control efforts, enabling precise and targeted interventions.

Thirdly, it facilitated the management of work, production, and school resumption. As the COVID-19 situation gradually improved, large-scale work, production, and school resumption across the country became inevitable. The health code enabled enterprises, schools, and other institutions to effectively monitor the health status of employees and students. This allowed targeted joint prevention and control measures to be implemented, preventing large-scale gatherings from triggering a resurgence of COVID-19 infections during reopening phases.

2.3.2 Enhancing Factor Organization Capability

Traditional social governance under normal circumstances is based on a bureaucratic hierarchy, emphasizing top-down policy implementation. In contrast, smart epidemic prevention represented by the health code in collaborative governance places greater emphasis on network connections, cross-departmental collaboration, and breaking organizational barriers, thereby enhancing factor organization capability. This capability played a critical role in the maturation and diffusion of the health code. One notable example is the establishment of the Data Resource Management Bureau at the provincial, municipal, and county (district) levels during Zhejiang Province's previous institutional reform. The district-level "Yuhang Green Code" could only access limited travel information, whereas, through municipal-level coordination, the "Hangzhou Health Code" could integrate roaming data from the three major telecommunications operators to determine whether individuals had passed through epidemic zones or stayed in other high-risk areas. This significantly enriched individuals' data profiles, with a simple upgrade aligning with core epidemiological principles. The Hangzhou Data Resource Management Bureau also collaborated with relevant business departments to connect previously segmented data systems, including confirmed and suspected case data managed by the Health Commission, contact tracing data managed by the transportation department, and migrant population data managed by the public security department.

The health code's development, launch, and execution process highlighted the strong coordination capability of Hangzhou's local government in governance. The local government played a leading role but did not monopolize public risk governance. Faced with the urgent challenges of epidemic prevention and the resumption of work and production, market entities, social organizations, and the public collaborated and interacted under government leadership. Through pragmatic optimization, collaborative innovation, and multilateral coordination, Hangzhou's governance measures demonstrated high value for replication and promotion.

2.3.3 Improving Digital Governance Levels

The health code, as an essential tool for collecting citizen information during epidemic prevention, served as an automated form-filling tool and online health certificate for individuals. However, for management institutions, it functioned as a Software-as-a-Service (SaaS) system capable of real-time tracking of residents' health conditions and monitoring key population travel trajectories. Previously, tracing individuals' travel histories required manually piecing together records from stacks of files, a process prone to errors and omissions. In the era of big data, the government utilized digital tools such as the health code to implement travel regulations in administrative regions. Every scan logged the user's geographic location, allowing management institutions to construct comprehensive, efficient travel trajectory profiles based on time and space sequences, facilitating subsequent tracing and management (Ministry of Civil Affairs of the People's Republic of China, 2020).

Additionally, the health code accelerated the digital transformation of infectious disease monitoring, healthcare, and related fields. It enhanced data interoperability in epidemic monitoring, material supply assurance, and other areas, reinforcing the data foundation for epidemic prevention. By promoting front-end digital collaboration and one-to-many data usage, it expedited multi-stakeholder coordination and multi-scenario application.

The health code also drove the expansion of digital innovation applications into work, healthcare, and other production and living scenarios, enriching data sources and dimensions. Leveraging the technical advantages of market entities, it encouraged platform enterprises, industry associations, and social organizations to explore new technologies and methods for data development. Through in-depth analysis of structured and unstructured data, it improved data utilization value and enhanced the scientific accuracy of risk identification and early warning systems.

2.3.4 Innovating Social Governance Mechanisms

In various digital governance scenarios, the health code has provided a new governance tool and digital technology platform for social governance. It has also innovated governance mechanisms for diverse social organizations, including cities, villages, communities, and enterprises, in their participation in public affairs. This has offered both a physical carrier and a virtual data space to support the realization of precise public services by the government. From departmental collaboration to the establishment of grassroots community defenses and balancing diverse goals, the health code has successfully addressed numerous challenges in public health management. By continuously exploring the data value of the health code, expanding application scenarios, and improving data protection measures, it has transitioned into a tool for routine governance, providing opportunities for the development of online public services and social governance, thereby adapting to the new trends in national governance modernization (Shan, 2021).

Moreover, the extension of the health code has provided a platform entry point for the transformation of digital government. By overlaying application scenarios and connecting more data, the health code has enabled the inclusion of most citizens into government platforms, facilitating the rise of responsive and service-oriented online governments. As an integrated digital profile of individuals, the health code maximizes the advantages of digital technology, integrating it as a new factor into traditional society to promote the positive transformation of economic and social structures. The iterative upgrade of the health code is rooted in the government's Internet of Things platform. Based on individuals, it ensures safety and privacy while integrating public service information, such as social security, health insurance, medical education, and transportation, with personal health status information. This promotes "multi-code integration", transforming the health code from a social management tool during the pandemic into a tool for digital public services in the post-pandemic era. The data, technical methods, and innovative thinking embodied by the health code have effectively served both the fight against the epidemic and the construction of daily social governance systems.

2.4 *Lessons Learned*

2.4.1 A People-Centered Approach as the Foundation of the Health Code

A people-centered and user-oriented mindset is the foundation and core principle of the health code. Issues such as "overlapping codes" and blanket policies in some regions stem from policymakers prioritizing administrative convenience over the fundamental principle of being people-centered. The primary purpose of the health code is to ensure public health safety during epidemic prevention, focusing on the precise identification and management of risks in high-risk areas. Policies that exceed the boundaries of maintaining public safety require comprehensive evaluation and careful consideration.

Firstly, the issue of the digital divide deserves attention. While health code travel offers low cost, broad coverage, high efficiency, and traceability, it has significant benefits for preventing gatherings, obtaining control data, and conducting epidemiological investigations. However, for older adults unfamiliar with smartphones, applying for health codes or scanning QR codes can be challenging. Forcing seniors to learn these tasks is neither flexible nor realistic. Reports of older adults being unable to enter supermarkets, shopping malls, or use public transportation due to their inability to scan codes, sometimes resulting in conflicts with epidemic prevention personnel, have been common.

Secondly, while artificial intelligence and big data outperform humans in many areas, it must always be remembered that technology exists to serve people. As a basic safeguard for human rights, manual processes should retain their fundamental place. Decision-making should not be entirely entrusted to machines; human review and real-time supervision must be strengthened to ensure continuous human involvement in decision-making processes and to prevent technological alienation. Therefore, only by continuously improving from the management perspective while adopting a citizen-centered perspective, particularly by balancing social governance with data security, can digital governance methods represented by the health code be effectively and appropriately utilized to achieve good governance.

2.4.2 Breaking Down Information Barriers as the Foundation of the Health Code

The government, as the primary force in social governance, possesses more information and resources than other stakeholders. In contrast, the public, enterprises, and social organizations, due to their inherent characteristics, have access to far less information. If there is a lack of communication and coordination between the government and other stakeholders, "information silos" and similar barriers will arise. Digital governance represents a transformation in governance philosophy. It is not merely about building a technical platform or utilizing a tool but is a critical means and instrument for achieving collaborative social governance. Hangzhou's "Digital Battle Against the Epidemic" demonstrates that digital technologies, by coordinating greater societal forces, can not only enhance administrative efficiency and effectiveness but also drive digital innovation in collaborative social governance systems.

To maximize the value of the government's data advantages, the key lies in achieving government data openness and information sharing. Digital governance, through internet technology, facilitates greater transparency of information, enabling fragmented data to be consolidated and shared across departments via a unified information platform.

First is to establish a proactive and standardized data openness mechanism. It is essential to define the scope of data openness. For emergencies, including major epidemics, the timeliness, accuracy, and completeness of data openness must be ensured, with data dynamically updated based on actual needs. Focusing on centralized data aggregation and open applications, regulations should be developed for data classification and grading standards, data openness platforms, and data quality management. This will expedite the opening and application of data related to public health emergencies to society.

Second is to build a government-business interactive data sharing mechanism. Information sharing reduces asymmetries between the government, enterprises, society, and the public, enabling societal forces to play a more significant role in governance and improving the alignment between social needs and governance capacity. Data should be allocated through market-driven mechanisms, where core tasks, such as data collection, storage, and platform construction during major emergencies, can be led by the government. Meanwhile, the development and utilization of data resources should be supported and encouraged through market participation to fully unlock the potential value of data.

Last is to clarify the roles and responsibilities of each stakeholder. The government should adopt a holistic perspective, strengthening relevant supporting policies and regulations. Enterprises should focus on leveraging their strengths in technology research, development, and application, innovating big data technologies, and providing diverse data products. The public, through the use of data products, can actively participate in emergency management, completing the feedback loop for data sharing mechanisms.

2.4.3 Ensuring Data Security as the Key to the Health Code

While promoting the opening of data resources to society, it is crucial to emphasize data security. For governments, this involves two aspects: first, leveraging big data platforms to enhance data co-construction and sharing for collaborative governance, and second, using these platforms to monitor data quality and regulate erroneous or false data to prevent social issues stemming from data misuse. To achieve this, a comprehensive big data security framework should be established, encompassing foundational system layers, data layers, and application layers. Multidimensional safeguards, including technical, procedural, and operational protections, must be implemented to ensure the safe application of big data and prevent data breaches during epidemic prevention efforts. Strengthening awareness of data security protection, promoting knowledge about personal privacy protection, and preventing data leaks are critical to sustaining the effectiveness of digital epidemic prevention measures. Digital governance methods like the health code must effectively protect personal privacy, manage data securely, and prevent the leakage or misuse of epidemic-related data. Strict measures to ensure citizens' data security, establish reasonable unified usage standards, and precisely limit the scope of use are crucial for the successful advancement of digital governance methods represented by the health code. Governments should adeptly utilize internet technologies to explore new governance methods, ensuring that information and data is truly "sourced from the people and served for the people".

Currently, the rollout of health codes across regions has involved mandatory measures, leaving users with no choice but to passively accept data collection and usage rules to comply with epidemic prevention efforts. Therefore, it is necessary to legally safeguard users' right to be informed by clearly communicating the scope of data collection, usage regulations, and timeframes associated with the health code. Reasonable information disclosure should be practiced to prevent the exposure of users' personal privacy, allowing them to use the health code with confidence. Additionally, there is no unified standard for health code usage across regions, leading to risks of data misuse or theft. A unified usage standard must be established, specifying that the health code should only be used in public places where checking health status is necessary, such as during work resumption or production recovery. Moreover, data usage timeframes for the health code must be defined. Once the epidemic ends, data should be sealed or destroyed to prevent prolonged or unauthorized usage.

References

Amap. (2020). *Traffic analysis report of major Chinese cities*. Retrieved December 31, 2021, from https://report.amap.com/download.do

China Youth Daily. (2019). From "solving traffic jams" to "urban governance": Hangzhou's City Brain coordinates 1.2 billion data entries daily. Retrieved December 31, 2021, from https://shareapp.cyol.com/cmsfile/News/201912/31/web310242.html

CNR. (2021). *National Health Commission: Nationwide health code "one-code access" essentially achieved*. Retrieved December 31, 2021, from https://doi.org/10.1016/J.TRD.2021.103148

Communist Party of China News Network. (2020). *Hangzhou: Making the city smarter and more intelligent*. Retrieved December 31, 2021, from http://cpc.people.com.cn/n1/2020/0617/c64387-31749528.html

Gao, P., & Xu, M. (2020). Lessons from Hangzhou's "Digital Battle Against the Epidemic": Empowering collaborative governance with data. *Social Governance, 8*, 53–58.

Hangzhou CPPCC News Network. (2021). *Are the "exit first, pay later" parking scenarios for over 75,000 parking spaces in Hangzhou working well?* Retrieved December 31, 2021, from https://www.hzzx.gov.cn/cshz/content/2021-04/11/content_7945285.htm

Hangzhou Municipal Government. (2019). *245 public medical institutions in Hangzhou now achieve "one-time payment" for medical visits*. Retrieved December 31, 2021, from http://www.hangzhou.gov.cn/art/2019/7/15/art_1228974663_36952697.html

Ministry of Civil Affairs of the People's Republic of China. (2020). *Research and improvement strategies on health codes*. Retrieved December 31, 2021, from https://xxzx.mca.gov.cn/article/xgzs/202004/20200400026407.shtml

People's Daily. (2019). [Innovative social governance] Hangzhou, Zhejiang: Building the "City Brain" to make cities smarter and life better. Retrieved December 31, 2021, from http://jl.people.com.cn/n2/2019/1120/c349771-33560089.html

People's Daily. (2020a). The iteration and upgrade of "digital" tourism playstyles. Retrieved December 31, 2021, from http://zj.people.com.cn/n2/2020/0116/c186327-33720829.html

People's Daily. (2020b). WHO calls for decisive action against the epidemic. Retrieved December 31, 2021, from http://paper.people.com.cn/rmrb/html/2020-03/11/nw.D110000renmrb_20200311_5-16.htm

Shan, Y. (2021). The legitimacy and improvement of health code applications. *Chinese Public Administration, 5*, 53–60. Retrieved December 31, 2021, from https://dysw.cnki.net/kcms/detail/detail.aspx?filename=ZXGL202105011&dbcode=CRJT_CJFD&dbname=CJFDLAST2021

Shi, C., & Ma, L. (2020a). Collaborative governance, technological innovation, and smart epidemic prevention: A case study of health codes. *Party & Government Research, 4*, 107–116. Retrieved December 31, 2021, from https://xueshu.baidu.com/usercenter/paper/show?paperid=1u230xd0js3k0ve0jk710ad0km667661

Shi, C., & Ma, L. (2020b). Internet enterprises boosting the construction of digital government: A case study of health codes and "Zhezheng Ding."*Study Forum, 8*, 50–55. Retrieved December 31, 2021, from https://xueshu.baidu.com/usercenter/paper/show?paperid=196g0j80dm0f0re0y13704t043429019

The Paper. (2020). *Health code: A health guarantee for mobility.* Retrieved December 31, 2021, from https://m.thepaper.cn/baijiahao_6877828

Tianmu News. (2020). *How did Zhejiang's City Brain contribute to epidemic prevention in 2020?* Retrieved December 31, 2021, from https://tm.zjol.com.cn/news.html?id=258710

Wang, H. (2020). Hangzhou Health Code: Local innovation in risk governance and suggestions for its expansion and improvement. *Zhejiang Academic Journal, 3*, 36–41.

Zhejiang Daily. (2020). *The "national version" of the health information code is here.* Retrieved December 31, 2021, from https://china.zjol.com.cn/gnxw/202003/t20200303_11734796_ext.shtml

Zhejiang News. (2019). *Experience the convenience of the "digital city": Hangzhou scenic spots hit new visitor records.* Retrieved December 31, 2021, from https://zj.zjol.com.cn/news.html?id=1192324

Zhejiang Online. (2019). *Hangzhou's "City Brain" brings convenience to all aspects of life.* Retrieved December 31, 2021, from https://biz.zjol.com.cn/zjjbd/cjxw_11149/201907/t20190711_10566127.shtml

Chapter 5
Case Studies on Data-Driven Urban Governance

1 Identifying Companion Travel Behavior Using Subway Card-Swipe Data

Companion travel behavior refers to the intentional act of two or more people traveling together. Such travelers often share social relationships, such as being family members, friends, colleagues, or fellow townspeople. In real life, companion travel behavior is ubiquitous—for instance, two people shopping, watching movies, attending sports events, traveling, studying, or commuting together. Existing research on walking behavior indicates that in some public spaces, such as large shopping malls, museums, and commercial streets, over half of the pedestrians choose to walk in pairs or groups (Moussaïd et al., 2010). Currently, most studies rely on collecting surveillance video information to manually identify companion travelers. Costa (2010) collected extensive pedestrian walking videos in historical districts and commercial streets in the Italian cities of Pescara and Bologna. By analyzing conditions such as visual contact, body orientation, gestures, and conversations among pedestrians, Costa identified companion travelers, ultimately finding 1,020 groups comprising approximately 2,500 companion travelers. While this approach achieves high accuracy for walking scenarios, its efficiency is very low, making it difficult to simultaneously identify large numbers of companion groups.

The case we present primarily attempts to rapidly and extensively identify potential companion travel behavior in public transit scenarios using subway card-swipe data. Taking subway station entry and exit as an example, when passengers swipe their cards to enter or exit the subway, their card-swipe time and location data are recorded. If two people are companion travelers, their subway station entries and exits will match, and their entry and exit times will be very close—usually within 1 to 3 s. Conversely, if two people's subway station entries and exits are identical and their times are very close, they are likely to be companion travelers (see Table 1). For strangers, maintaining very close entry or exit timing is relatively easy. However,

W. Zhang and Y. Zhang, *Digital Urban Governance in China*, Qizhen Humanities and Social Sciences Library, https://doi.org/10.1007/978-981-96-6325-5_5

in the highly complex travel scenarios of subway systems, the probability of maintaining close proximity at both entry and exit significantly decreases. Various factors influence the spatiotemporal proximity of strangers' card-swipe behaviors, such as differing destinations, peak commuting times, subway car space constraints, and the spatial layout of subway stations.

To test the feasibility of using subway card-swipe data to identify companion travel behavior, we conducted exploratory experiments in Beijing. We used one week of Beijing subway card-swipe data from April 5 to April 11, 2010, comprising 15.2 million travel records. We set a 3-s threshold for temporal proximity to identify companion travel behavior. Specifically, if two trip records had the same start and end stations, and both the start time and end time intervals between the two trips were less than 3 s, these records were identified as companion travel records. Out of the 15.2 million travel records, we identified 431,500 companion travel records, accounting for approximately 2.84% of the total, involving 212,000 groups.

Figure 1 illustrates the temporal pattern of companion travel behavior on a daily basis. The companion travel ratio refers to the proportion of group trips to the total number of trips on a given day. The Beijing subway card-swipe data indicated whether a card was a student card. Based on this information, Fig. 1 shows the behavioral characteristics of three categories: all cardholders ("overall"), all student cards ("students"), and all non-student cardholders ("others"). For the "overall" distribution, Monday, April 5, fell on the Qingming Festival holiday, showing the highest companion travel ratio (4.34%). This aligns with common expectations, as many residents choose to go outdoors for spring outings, visit family, or pay respects at ancestral graves during the Qingming Festival. The second-highest companion travel ratios occurred over the weekend, with Saturday (3.86%) slightly higher than Sunday (3.31%). This trend reflects people's preference for leisure activities on Saturdays, as they may opt to rest at home on Sunday in preparation for the workweek. The companion travel ratio during weekdays was significantly lower than on holidays and weekends. The "overall", "students", and "others" categories all exhibited similar characteristics: cardholders were more likely to engage in companion travel on holidays and weekends. Additionally, students consistently showed a significantly higher companion travel ratio than others on any given day, reflecting their preference for traveling in groups. This could include commuting to and from school with classmates or participating in activities organized by schools or extracurricular institutions.

Figure 2 illustrates the temporal distribution pattern of companion travel behavior within a single day. The results show that the afternoon has a higher proportion of companion travel, with the ratios at 14:00 and 15:00 reaching 3.42% and 3.38%,

Table 1 Examples of Metro Card Swipe records for group travelers in Beijing

Anonymized card no.	Date	Entry time	Entry station name	Exit time	Exit station name
1823504857	2010/4/8	10:30:18	Tian'anmen West	11:00:30	Beijing Zoo
1827743210	2010/4/8	10:30:19	Tian'anmen West	11:00:28	Beijing Zoo

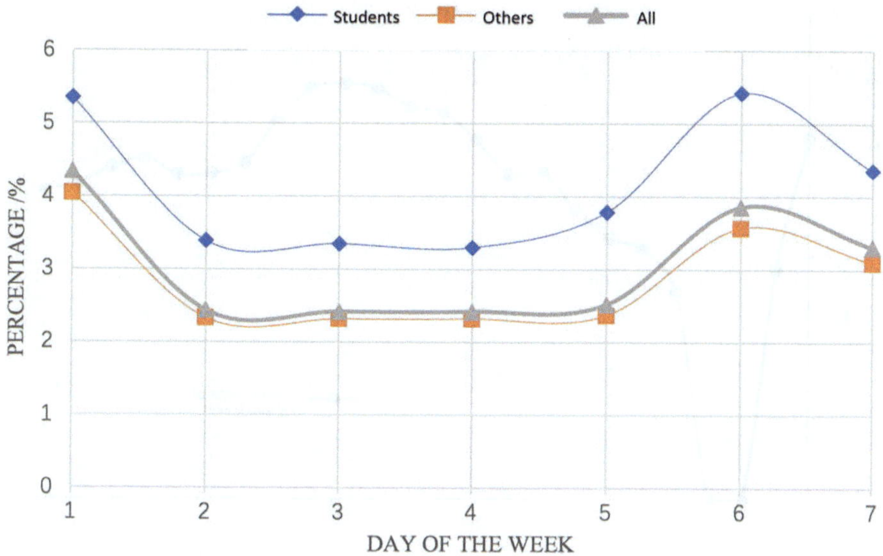

Fig. 1 The distribution pattern (day of the week) of group travels in Beijing. *Note* The horizontal axis represents the period from Monday (April 5) to Sunday (April 11), 2010

respectively. Additionally, 20:00 also exhibits a relatively high companion travel ratio. The higher companion travel ratios in the afternoon and evening reflect people's willingness to engage in leisure and social activities during these times. Conversely, the companion travel ratios during the morning and evening rush hours are relatively low, indicating that most trips during these periods are individual, primarily for commuting purposes. Although the number of trips during peak hours is significantly higher, which technically increases the likelihood of mistakenly identifying individuals who coincidentally swipe their cards at similar times as companion travelers, the results still show relatively lower companion travel ratios during these times. This indicates, to some extent, that the proposed method effectively distinguishes between individual and companion travel behaviors.

To analyze the spatial distribution pattern of companion travel, we examined the top 10 subway stations with the highest companion travel ratios. Table 2 shows that the Olympic Sports Center subway station has the highest companion travel ratio (6.13%), while Wangfujing station, ranked 10th, has a ratio of 3.86%. On the map, the top 10 stations can be categorized into two cluster hotspots and two standalone hotspots. The two cluster hotspots are the Olympic Park cluster and the Forbidden City cluster. The two standalone hotspots are Beigongmen and Zoo subway stations. Beigongmen is a station on Beijing Metro Line 4, located on the east side of Beigongmen Road in Haidian District, Beijing, near the north gate of the Summer Palace. The station is directly adjacent to the Summer Palace scenic area and the Central Party School, and its surroundings include attractions such as the Old Summer Palace, Beijing University, and the University of International Relations. A

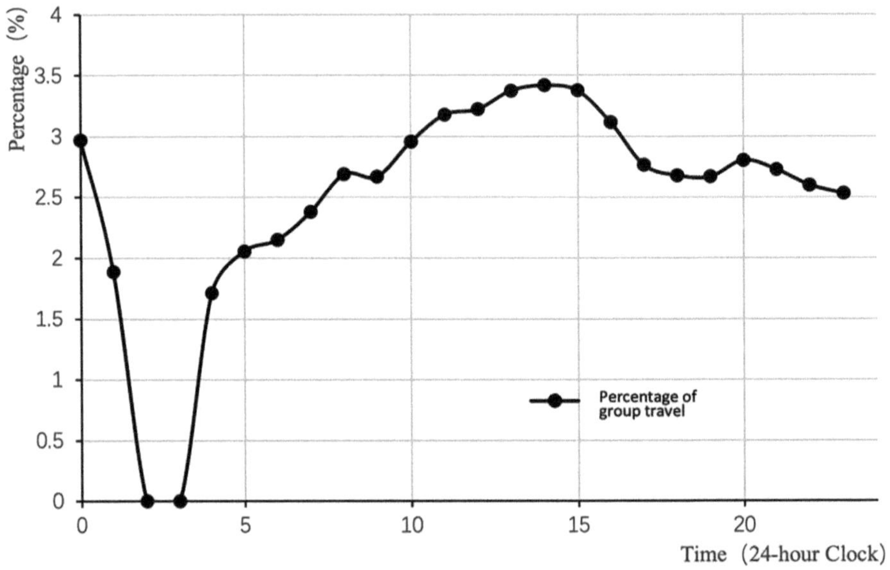

Fig. 2 The time distribution pattern (in hours) of group travel

significant portion of passengers entering and exiting Beigongmen station are tourists and student groups with a preference for companion travel. Similarly, the area around Zoo station includes scenic spots and schools like those near Beigongmen (e.g., Beijing Zoo, Beijing University of Civil Engineering and Architecture, and Beijing No. 56 Middle School). In addition, it features major exhibition resources such as the Beijing Exhibition Center and the Beijing Planetarium.

In the Beijing Olympic Park area, there are numerous subway stations with high proportions of companion travel, which is closely related to the presence of large

Table 2 Top 10 Metro stations witnessing group travel in Beijing

Rank by group travel proportion	Station name	Group travel proportion (%)
1	Olympic Sports Center	6.13
2	South Gate of Forest Park	6.09
3	Olympic Park	5.65
4	Xidan (M69)	5.02
5	Tian'anmen West	4.68
6	Beijing Zoo	4.63
7	Tian'anmen East	4.58
8	Xidan (M14)	4.36
9	Beigongmen	4.35
10	Wangfujing	3.86

public facilities in the vicinity. These facilities serve diverse purposes, including tourism, shopping, sports events, exhibitions, conferences, and outdoor activities. Located in Chaoyang District, Beijing, the Olympic Park was the venue for the 2008 Beijing Olympics and Paralympics. After the Olympics, it evolved into a comprehensive public activity center combining sports events, exhibition centers, cultural and educational functions, leisure, and shopping. The Forbidden City cluster mainly includes two categories of facilities that attract group activities: high-quality tourist and cultural facilities, such as the Forbidden City, the National Grand Theater, the National Museum, and Tiananmen Square; and major commercial shopping centers like Xidan and Wangfujing. The examples of the ten subway stations indicate that companion travel behavior is more likely to occur near subway stations close to leisure, tourism, and shopping facilities.

Research on companion travel behavior has potential applications in areas such as public transit group ticket policies, public event management, and public space design. For example, large public events often attract significant crowds, potentially leading to congestion or unforeseen incidents like stampedes. Companion travelers, who tend to stay close to other group members, can affect crowd movement, slowing down some individuals and disrupting overall pedestrian flow. Thus, guiding companion travelers' behavior reasonably and orderly is crucial for effective public event management. Moreover, the findings reveal which subway stations, lines, and surrounding areas attract more companion travelers. Based on these insights, public spaces can be consciously designed to better accommodate companion travelers. For instance, subway waiting areas can include more side-by-side seating arrangements to facilitate social interactions, display advertisements and information about leisure and entertainment activities, and offer public safety messages, creating a more companion-friendly urban environment.

Below, we provide a more systematic introduction to the group ticket policy in public transportation. A group ticket refers to a type of fare designed for two or more passengers traveling together. Currently, in the public transportation systems of developed countries, group tickets have become an effective complement to single tickets and are widely adopted in many cities (e.g., Seoul, Paris, London, New York, and Chicago). The diversity of group ticket types helps attract more people to choose public transportation, especially those with group travel needs. Transit companies may benefit from increased ridership, while passengers can reduce their travel costs. Moreover, higher utilization rates of public transportation contribute to reducing traffic pollution and building environmentally friendly societies. Additionally, group tickets offer social benefits not available with single tickets, as they directly or indirectly encourage social interaction, enhance travel safety, increase journey enjoyment, and protect or support vulnerable groups such as students and the elderly. They play a positive role in maintaining positive social relationships (e.g., family bonds) and building harmonious societies. Thus, implementing group ticket policies is conducive to sustainable urban development, benefiting not only environmental and economic dimensions but also social dimensions of sustainability.

However, the development, implementation, and optimization of group ticket policies currently face two major issues. First, the formulation of group ticket policies often relies on policymakers' subjective experience and lacks systematic understanding and analysis based on objective data. In fact, a search on Google Scholar for the term "transport group ticket" did not reveal any academic papers systematically discussing group ticket policies. Second, group tickets are typically purchased separately and require manual ticket inspection for station entry and exit. This increases labor costs, reduces efficiency, and significantly limits the implementation of group tickets.

The research presented in this case study can provide some insights into addressing these issues. First, the analysis of companion travel behavior based on subway card-swipe big data helps us better understand the spatial and temporal characteristics, scale, and potential economic benefits of companion travel. Based on these characteristics, we can further propose optimization strategies for group ticket policies. Depending on group size, category, implementation time, and route, we can suggest various types of group tickets, such as general group tickets, family tickets, companion tickets, off-peak group tickets, weekend group tickets, and long-distance commuting group tickets. For example, our research shows that students have a preference for companion travel. Introducing group tickets specifically targeting student groups could meet their needs, promote social interaction during their journeys, and enhance their travel safety to some extent.

Secondly, traditionally, one transit card corresponds to a reusable single ticket. This single ticket is processed automatically by smart card readers installed at the entrances and exits of subway or bus stations. The fare calculation typically considers factors such as the card type (e.g., student card), travel time, distance, and relevant policies. Here, we propose an automatic ticketing and pricing system that enables one transit card to correspond to a reusable group ticket. Taking buses as an example, the basic logic for automatic ticketing and pricing is as follows: "If several passengers swipe their cards in close sequence when boarding and alighting, they qualify for the corresponding group ticket discount." Essentially, this logic aligns with the identification method for potential group ticket users. A similar example can be observed in Taipei City. The "Love Card" is a transit card provided for individuals with disabilities in Taipei. Holders of a Love Card can apply for a "Love Companion Card" for one necessary companion (see Fig. 3). When used independently, the Love Companion Card is charged at the full fare. However, if it is used immediately after the Love Card, it qualifies for a 50% discount. This demonstrates that the Taipei Love Companion Card can be viewed as a reusable group ticket, and its identification logic fundamentally aligns with the principles discussed in this book.

This case study provides a simplified method for identifying companion travel behavior and explores the spatial and temporal characteristics of such behavior. The method effectively distinguishes between companion and individual travel behavior but has certain limitations. For instance, using a fixed 3-s threshold may exclude some companion travelers with longer card-swipe time intervals. Additionally, the method cannot identify some types of companion travel behavior, such as travelers who share the same starting station but not the same destination station. For further

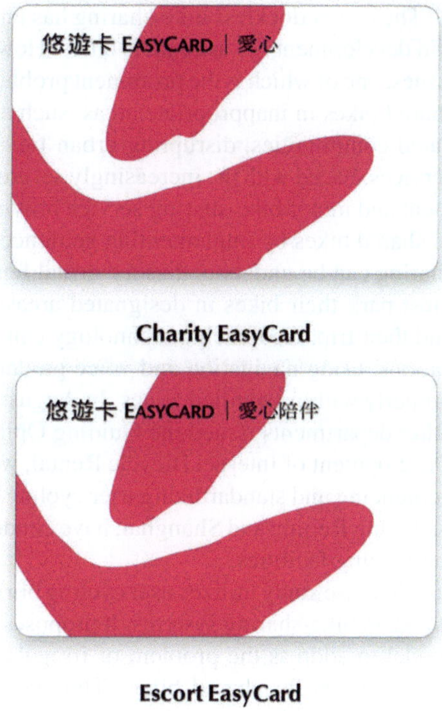

Fig. 3 Taipei EasyCard (Charity and Escort). *Source* https://www.easycard.com.tw

discussion and refinement of the method, refer to Zhang and Martens (2018) and Zhang et al. (2018, 2024b).

2 Using User Cycling Big Data to Support Geofencing Planning

Bike-sharing is a service that allows the general public to share the use of bicycles, gaining popularity worldwide since the 1960s. Bike-sharing services bring numerous social, economic, and environmental benefits, such as promoting public transportation, alleviating traffic congestion, reducing and controlling diseases, lowering greenhouse gas emissions, and eliminating noise pollution (Zhang et al., 2023, 2024a). In 2015, a new generation of internet-based dockless bike-sharing services emerged and quickly rose in popularity in China. Compared to traditional docked bike-sharing systems, dockless bike-sharing achieved a technological breakthrough by allowing bikes to be "used and parked anywhere". This unprecedented convenience made dockless bike-sharing widely popular, enabling rapid market expansion in a short period. It created new job opportunities and boosted urban socio-economic development while providing convenient travel services.

The rise of dockless bike-sharing has undoubtedly contributed to the construction and development of sustainable cities. However, it has also brought about new urban issues, one of which is the prominent problem of "irregular parking". Many users park shared bikes in inappropriate areas, such as sidewalks, subway entrances, or within gated communities, disrupting urban functions and the proper use of bike-sharing services. Faced with the increasingly severe problem of irregular parking, the government and major bike-sharing service providers plan to promote the standardized use of shared bikes by implementing geofenced parking facilities. Geofencing for bike-sharing can be understood as a "virtual fence" without physical form. Ideally, users must park their bikes in designated areas; otherwise, they cannot lock the bike to end their trip. Geofencing technology can integrate with bike-sharing mobile applications, using navigation and voice guidance within the app to direct users to park properly within specified zones. In August 2017, the Ministry of Transport and nine other departments issued the Guiding Opinions on Encouraging and Regulating the Development of Internet Bicycle Rental, which included policy guidance supporting geofencing and standardizing user cycling behavior. Since early 2017, several cities, including Beijing and Shanghai, have conducted small-scale pilot tests of shared bike geofencing facilities.

This case study utilizes user cycling big data to support the geofencing planning of dockless bike-sharing systems. It proposes a method based on the location allocation model to address the problem of irregular parking and optimize the spatial layout of geofences for shared bikes. The cycling data used in the study was provided and authorized by a shared mobility company, with full consideration given to user privacy protection. The dataset includes approximately 777,900 randomly sampled cycling records in Shanghai from September 15 to 30, 2017, generated by 299,000 shared bikes. Each record contains information such as trip ID, anonymized user ID, bike ID, trip start (end) time, and the latitude and longitude of the trip's start (end) points. This large dataset, which is detailed and spans an extended period, provides an objective reflection of the cycling behavior of shared bike users.

Firstly, the study preprocesses raw data from OpenStreetMap street segments and shared bike trip records to derive Shanghai's road network, a 50-m-by-50-m city grid, the spatial distribution of shared bike origin-destination points within the city grid, and candidate locations for geofences (represented by the grid cells). Secondly, four scenarios with different numbers of geofences are established, and a location allocation model is used to select specific geofence locations from the candidate grid cells based on parking demand.

The location allocation model is the primary method applied in this case study. Commonly used to solve siting problems for public facilities or organizations, the model has been widely applied in areas such as the location selection of healthcare facilities, fire stations, and schools (Rahman & Smith, 2000). The location allocation model determines facility points by considering the location and demand characteristics of demand points and assigns the demand points to respective facility points to meet the needs of the demand points most efficiently. The allocation method depends on the type of problem being addressed; common types include minimize impedance (minimizing the total distance from demand points to selected facility points), maximize coverage (satisfying the greatest demand within the coverage area) and minimize facilities (minimizing the number of facilities needed to cover demand points), etc. The objective of this study is to plan geofence facilities for shared bikes to accommodate parking demand in their vicinity, making the location allocation model suitable. To maximize coverage of shared bike parking demand, the study selects "maximize coverage" as the problem type.

Parking demand within a grid cell is defined by the number of origin–destination (OD) points. Candidate grid cells for geofences are those with parking demand exceeding five, meaning the grid contains more than five OD points.

In central Shanghai, which has a high road network density, the city is divided into 2,718,296 50 m × 50 m urban grids. Among these, 186,881 grids (approximately 6.87%) show demand for shared bike parking, while over 90% of the grids have no such demand. For grids with parking demand, the average demand is 8.3 instances, with a standard deviation of 18.1. Most of the parking demand is concentrated within the Outer Ring Expressway of Shanghai, particularly in central districts such as Huangpu, Jing'an, and Yangpu. Due to the lack of data reflecting institutional frameworks, urban infrastructure, and public preferences, the current study determines geofence locations solely based on parking demand derived from shared bike user cycling data. The study sets the threshold for parking demand within candidate geofence grids at 5. Based on this standard, 58,941 urban grids are identified as candidate geofence locations, with an average parking demand of 22.1 instances per grid and a standard deviation of 27.5. Most candidate geofence locations are within the Outer Ring Expressway and concentrated in central districts.

The distribution of geofence locations under four planning scenarios is then examined. When the number of geofences is set at 5,000, most geofences are concentrated in central districts and Pudong New Area. As the number of geofences increases, geofences quickly expand from the central districts to the suburbs, though most geofences remain within the Outer Ring Expressway. Table 3 shows the percentage of shared bike parking demand accommodated in Shanghai and its districts under the four geofence planning scenarios. The table reveals that as the number of geofences increases, the marginal increase in accommodated parking demand decreases. For instance, when the number of geofences is 5,000, 85.0% of shared bike parking demand can be met. Increasing the number of geofences to 7,500, 10,000, or 12,500 results in additional demand coverage of 6.8, 3.2, and 1.3 percentage points, respectively. Table 3 also indicates spatial disparities in the coverage of parking demand across districts under different planning scenarios. In central districts such as Huangpu, Xuhui, Changning, Jing'an, Putuo, Hongkou, and Yangpu, over 90% of parking demand is satisfied across all scenarios. In contrast, suburban districts like Qingpu and Chongming have significantly lower coverage. Jinshan and Fengxian districts perform the worst; even with 12,500 geofences, less than 30% of parking demand in these areas is met. Considering that when the number of geofences reaches 10,000, over 90% of shared bike parking demand is satisfied and each district has geofence facilities, this study identifies the 10,000-geofence scenario as the optimal geofencing plan. This approach ensures relatively high demand coverage with a reasonable number of geofences.

Table 3 Proportions of parking needs met by electronic fences under different planning scenarios

District	Electronic fence planning scenario							
	5,000		7,500		10,000		12,500	
	Number of fences / unit	Percentage of demand met / %	Number of fences / unit	Percentage of demand met / %	Number of fences / unit	Percentage of demand met / %	Number of fences / unit	Percentage of demand met / %
Shanghai	5,000	85	7,500	92	10,000	95	12,500	96
Huangpu	267	95	357	98	428	99	528	100
Xuhui	440	94	576	98	711	99	841	99
Changning	277	93	360	97	444	99	528	100
Jing'an	376	95	474	97	564	99	666	100
Putuo	382	92	523	96	645	99	767	99
Hongkou	263	96	328	99	379	100	436	100
Yangpu	468	94	627	98	753	99	883	99
Minhang	551	74	961	86	1,340	91	1,709	94
Baoshan	514	84	778	92	1,028	96	1,249	97
Jiading	152	63	287	74	468	81	651	85
Pudong New Area	1,,081	77	1,720	87	2417	92	3,142	94
Jinshan	0	0	0	0	1	3	7	13
Songjiang	203	56	421	73	656	83	848	86
Qingpu	18	24	58	42	109	53	162	59
Fengxian	1	2	8	12	22	21	39	27
Chongming	7	28	22	56	35	67	44	71

To mitigate the issue of "irregular parking" associated with dockless shared bikes, this book proposes leveraging user cycling big data and the location allocation model to support the spatial layout planning of geofences for shared bikes. Using Shanghai as a case study, the research analyzed shared bike user cycling big data from various districts of Shanghai from September 15 to September 30, 2017. The location allocation model was applied to calculate geofence layouts under four scenarios with different numbers of geofences, and the number of geofence locations accommodating different levels of parking demand was determined for the optimal scenario (10,000 geofence points). The research results indicate: (1) In all planning scenarios, most geofence locations are concentrated in the central districts; (2) As the number of geofences increases, geofence locations expand from the central districts to suburban areas; (3) In all scenarios, over 85% of the city's parking demand is met; (4) There are spatial differences in parking demand coverage across the city, with central districts accommodating significantly more demand than suburban areas. (5) In the optimal scenario, 94% of parking demand can be met, and theoretically, all shared bikes in Shanghai can be accommodated within geofences.

Due to the relatively short history of shared bikes and the difficulty of obtaining cycling data, most existing research has focused on discussing policy optimization for shared bike operations and management, with few conducting quantitative scientific studies. This study utilized a large and rich dataset provided by a shared mobility operator, offering a strong foundation for precise quantitative research. It is the first detailed quantitative analysis of parking issues associated with dockless shared bikes, presenting a novel research perspective.

The findings of this study provide a rational and effective reference for geofence planning, aiding governments and shared bike service providers in managing bike deployment and promoting standardized parking. This supports sustainable transportation models and facilitates the planning, development, and governance of sustainable cities. Notably, the geofence planning method proposed in this book does not yet consider constraints such as policy frameworks, urban infrastructure, and public preferences. In the future, incorporating these factors into the model could improve its accuracy and practicality, thereby supporting the development of more reasonable geofence planning schemes. Based on the results of this study, the following policy recommendations are proposed: firstly, governments and service providers can utilize the study's model to analyze usage data for shared bikes and update parking demand across urban areas; secondly, governments and service providers should collaborate in planning geofences. Thirdly, shared bike service providers need to update their technology accordingly, including providing real-time parking location information to users through mobile apps and designing reward and penalty mechanisms to encourage proper parking behavior. Fourthly, collaboration and communication among shared bike service providers should be encouraged for joint planning, usage, and management of shared geofences. Further discussion and refinement of geofence-related cases can be found in Zhang et al. (2019).

3 Using Mobile Signaling Big Data to Identify Individual Movement Trajectories

Mobile signaling data refers to information generated during the interaction between mobile devices and wireless communication networks. Spatial positioning methods for mobile signaling data primarily include cell-based positioning and measurement report (MR)-based positioning. Among them, cell-based positioning is computationally simple. It identifies the communication cell ID where the mobile device interacts and locates the user within the coverage area of that cell. The positioning accuracy is approximately 200–500 m in urban areas and 400–1000 m in suburban areas. MR-based positioning involves a slightly higher computational complexity but achieves an accuracy of 50–200 m.

With the widespread adoption of mobile phones and the popularity of mobile internet in recent years, the volume of mobile signaling data has increased dramatically. This data effectively captures resident behavior and has characteristics such as massive scale, diverse types, and dynamic spatiotemporal attributes. It overcomes the limitations of traditional static and outdated data, enabling real-time spatial monitoring of human activity. This makes it more suitable for measuring demographic characteristics. For instance, the number of active mobile users can be used to measure the intensity of human activities and the dynamic distribution of populations across different regions and time periods; long-term monitoring and analysis can help identify users' residential and workplace locations, uncover patterns of population mobility, and analyze the relationship between human activity intensity and urban planning, layout, and development. These insights can serve as references for the planning and construction of urban infrastructure. Several specific application cases are provided below.

3.1 Assessing the Vitality of Characteristic Towns

As an important component of new urbanization construction, characteristic towns emphasize on the one hand, industrial support, respecting market logic to the greatest extent and fully leveraging the agglomeration effect; on the other hand, enhancing policy supply to alleviate institutional, systemic, and resource constraints faced during development, achieving the integration of high-end elements within the town and the free flow of elements beyond its boundaries. Based on mobile signaling big data provided by a telecommunications operator in Zhejiang Province, the following case constructs a population and activity evaluation model to comprehensively analyze the population size, demographic labels, and activity patterns of 114 characteristic towns listed in the first three provincial-level creation batches in Zhejiang Province. This analysis evaluates whether the development of characteristic towns is stable, healthy, and sustainable. Measuring the vitality level of these towns can reflect the overall effectiveness of this policy implementation.

Specifically, based on mobile signaling data, the analysis identifies three categories of populations: "working population", "visiting population" and "resident population". Combined with POI data from Amap and enterprise registration information from the Administration for Industry and Commerce, it evaluates the completeness of supporting facilities, the vitality of community functions, and the status of industrial development in the characteristic towns.

By analyzing user activity trajectories, the working population within and around the towns can be identified (defined as individuals staying during daytime working hours for more than 10 days, with the highest frequency). This allows the evaluation of the absolute number of job opportunities provided within the towns and the relative number of job opportunities available within their surrounding areas. Figure 4's scatter plot of working populations in four quadrants illustrates the performance of various towns in terms of working population size and concentration (working population concentration = working population within the planned town area ÷ working population within a 1-km radius of the town area). The average working population size of characteristic towns is approximately 3,700, and the average working population concentration is 0.22, meaning these towns provide 22% of the job opportunities within their planned areas and a 1-km radius.

Based on mobile user registration information, place of residence, and mobile usage patterns, the relative changes in the proportions of various demographic groups (e.g., education level, income level, age) in the working population of characteristic towns can be compared with the corresponding proportions of the resident population in their respective districts or counties. This comparison provides a demographic profile of characteristic towns (see Figs. 5, 6, and 7) and measures their attractiveness

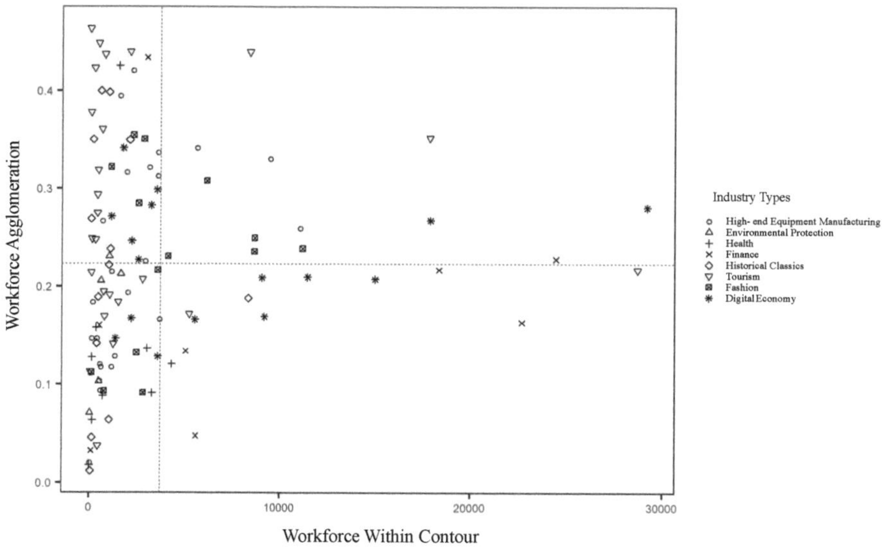

Fig. 4 Scatter plot of workforce size and agglomeration in four quadrants

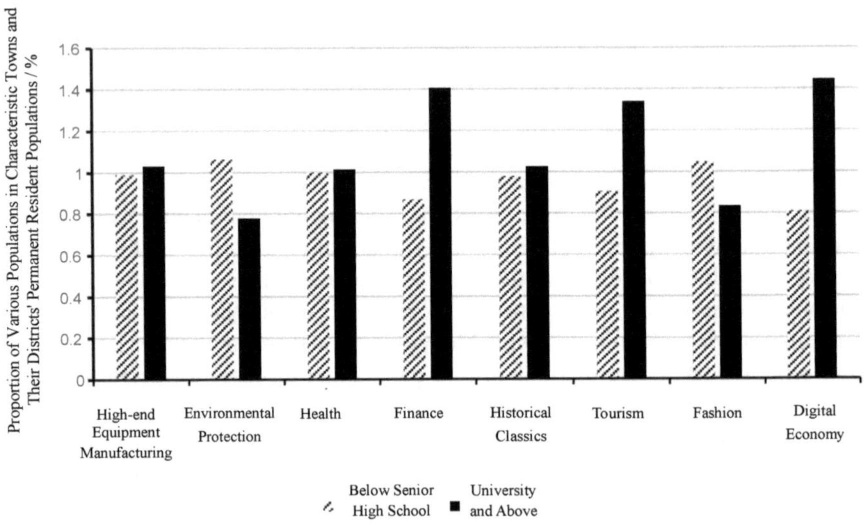

Fig. 5 Educational level group image

to different working groups. Overall, compared to the average levels in their respective districts or counties, characteristic towns attract a higher proportion of highly educated and younger working populations, offering higher incomes. Towns specializing in healthcare, finance, tourism, and the digital economy have a particularly high proportion of high-income workers. Towns focused on advanced equipment manufacturing, finance, and the digital economy show a particularly pronounced trend of attracting younger working populations.

Through long-term comparisons, it is possible to observe population inflow and outflow trends in characteristic towns over time. As shown in Fig. 8, apart from May 2019 and May 2020, the working population in the first two batches of characteristic towns in Zhejiang Province experienced a net inflow, displaying certain cyclical patterns. Among them, tourism-oriented and historically significant characteristic towns maintained relatively strong working population inflows in the second half of 2018. Meanwhile, environmentally focused characteristic towns from the second batch showed relatively strong working population inflows in the second half of 2019 and the first half of 2020.

By comparing mobile users' places of work and residence, the commuting distances and jobs-to-housing balance can be analyzed. Table 4 categorizes towns with poor commuting conditions into three types: the first type is towns with generally low commuting levels across the region. For instance, Dinghai Ocean Fishing Town in Zhoushan has short commuting distances for most of its working population, yet the optimization of commuting is poor. Second type is towns located far from major urban centers with dual commuting peaks. For example, Yiwu Silk Road Finance Town and Xinchang Intelligent Equipment Town show two commuting peaks: within 5 km and between 20–50 km. This indicates that these towns are extremely distant from

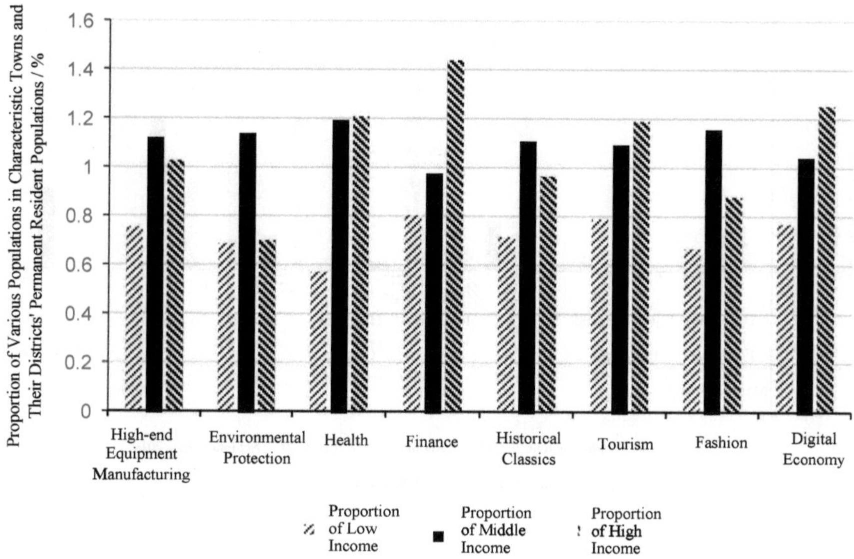

Fig. 6 Income level population image

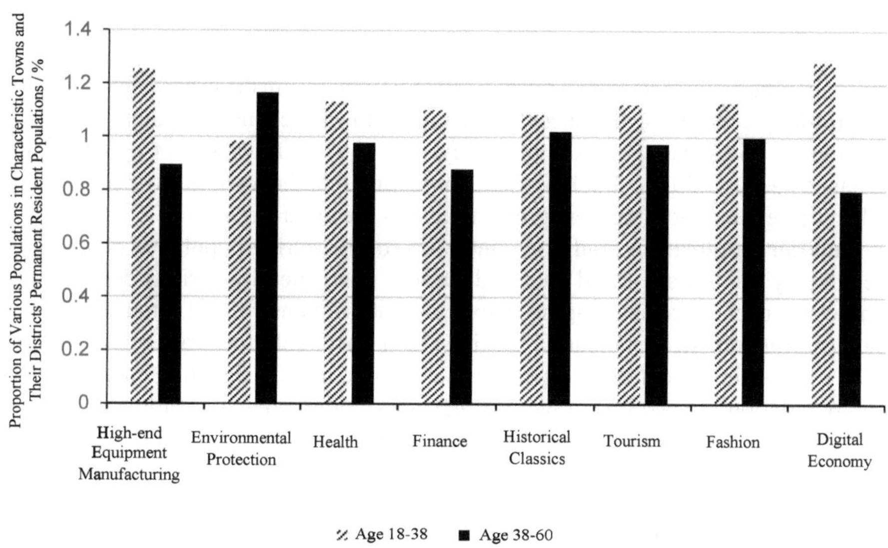

Fig. 7 Age structure population image

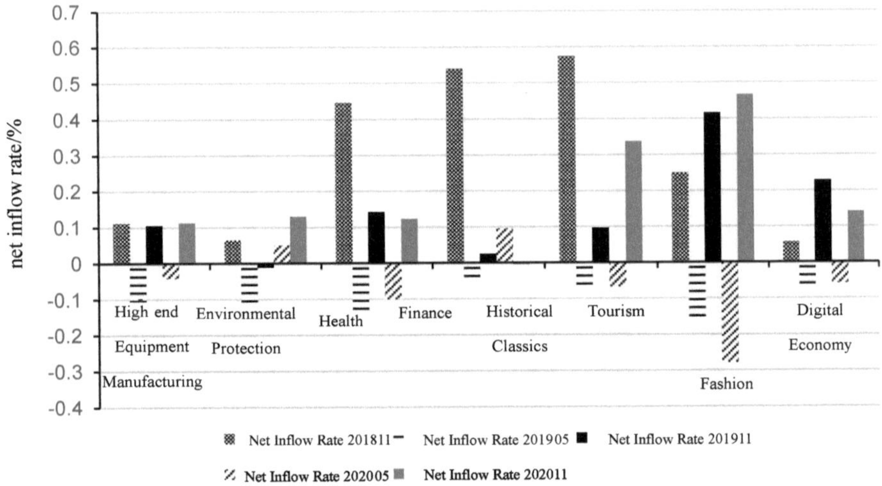

(a)

Fig. 8 Inflow rate of workforce in the first, second, and third batches of provincial characteristic towns. (a) First batch of characteristic towns; (b) Second batch of characteristic towns; (c) Third batch of characteristic towns

the main urban area, and some workers still choose to live in the urban center, leading to ultra-long commuting distances. Third type is towns near urban centers with short commuting distances. For example, Wuyi Hot Spring Town shows a commuting distance peak within the 3–5 km range, indicating that the town is close to the main urban center, and most workers prefer to live in the urban area.

Further Application: traditional methods for measuring population characteristics mostly rely on statistical data and surveys conducted by various departments. These approaches are challenging to execute, labor-intensive, and often suffer from inconsistent statistical standards, making them unsuitable for large-scale dynamic analysis. By simply defining a specific spatial area, mobile signaling data can identify population characteristics within that area, minimizing the influence of data collectors on metric measurement and ensuring authenticity and objectivity. Furthermore, based on this population information, cities can interpret economic development and regional characteristics through dimensions such as population, industry, and supporting facilities. This information can actively contribute to balancing jobs and housing within the area and improving the layout of supporting infrastructure. Additionally, data on target population characteristics and behavioral trajectories can uncover deeper group demands, facilitating the advancement and optimization of projects such as commercial facilities and real estate development.

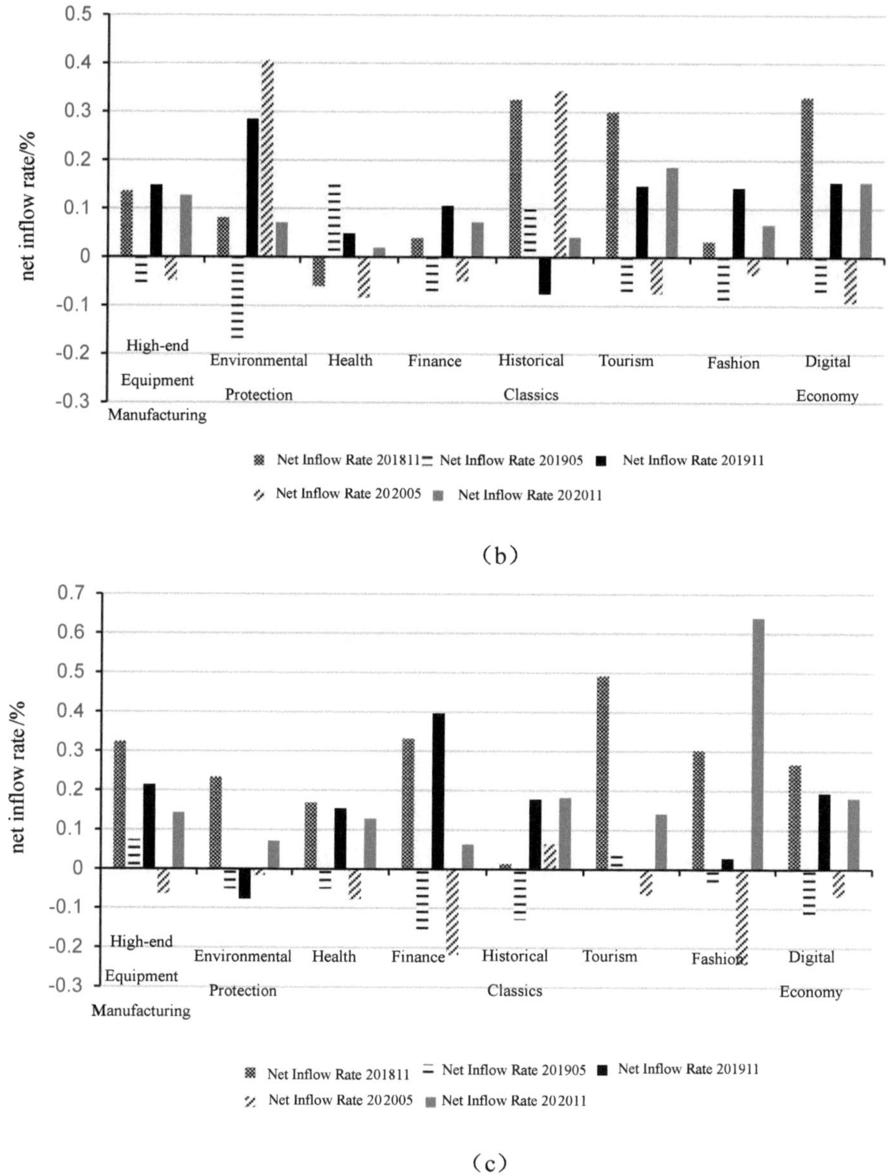

Fig. 8 (continued)

Table 4 Commuting situation in some towns

	0–1 km	1–2 km	2–3 km	3–5 km	5–10 km	10–0 km	20–50 km	50+ km
Unit: Number of people								
Dinghai Yuanyang Fishery Town	551	40	14	10	34	169	91	16
Yiwu Silk Road Finance Town	124	154	259	99	129	72	230	84
Wuyi Hot Spring Town	61	55	110	1210	99	105	3	180
Xihu Yunxi Town	67	34	183	760	133	182	20	113
Xinchang Intelligent Equipment Town	572	555	711	311	1406	242	1105	961
Dongyang: Woodcarving Town	16	16	29	31	28	41	30	23
Hengzhou: Circular Economy Town	30	134	75	75	85	166	1044	96

3.2 *Improving Public Transportation Efficiency*

By identifying the movement trajectories of mobile users, massive mobile signaling data can also empower the development of transportation systems. Recognizing individual travel behaviors and modes enables precise analysis and scientific management of urban transportation.

In transportation planning, the application of mobile signaling primarily relies on stored location information over a specific time period. By processing raw mobile signaling data, origin-destination (OD) points of mobile users within a given time frame can be extracted, or their travel time, distance, and trajectory can be determined. Additionally, with machine learning and map navigation route comparisons, travel modes can be identified. Compared to traditional traffic surveys for obtaining OD data, mobile signaling data not only saves significant human and material resources but also enhances precision and completeness. Mobile signaling data can identify, in real-time, the traffic flow and connectivity between various bus and subway stations.

Moreover, mobile signaling data offers advantages such as high frequency, objectivity, and comprehensive spatiotemporal coverage of travel. It enables the identification of individuals' complete activity trajectories, treating various modes of transportation (e.g., subway, bus, taxi, cycling, driving, and walking) as an integrated system. This facilitates exploring public transportation solutions, integrating travel resources, providing precise guidance, and creating a high-quality transportation system.

As an essential urban gateway, the distribution of passenger flow and the orderly planning of transportation resources at transit hubs play a critical role. However, mismatches in the supply and demand of various transportation resources often result in resource underutilization or passenger congestion. Mobile signaling data, with its real-time, dynamic, and comprehensive coverage, enables in-depth identification and analysis of user origin distributions, dwell times, key road traffic flows, and the ingress and egress relationships at transportation hubs. It can also provide solutions to potential problems.

For example, a telecommunications operator in Zhejiang Province analyzed passenger arrival and departure flows at Hangzhou East Railway Station using user trajectory data, combined with data from metro base stations, bus GPS systems, and hotel POI data. The analysis revealed that Hangzhou East Station handles an average daily passenger arrival volume of 135,000 people. With a daily passenger arrival volume of approximately 240,000 within Hangzhou city, Hangzhou East Station accommodates 56% of the city's total passenger flow. The primary destinations of arriving passengers include the West Lake Scenic Area, Xiaoshan International Airport, Yuhang High-Speed Railway Station, Hangzhou Coach Terminal, and West Coach Station. Therefore, a significant portion of arriving passengers transfers to other transportation hubs.

Figure 9 illustrates the time-variant characteristics of daily passenger arrivals based on historical data analysis. The peak period for passenger arrivals occurs between 10:00 and 20:00. Specifically, passenger arrivals surge from 12:00, peaking between 15:00 and 20:00. Meanwhile, the departure passenger volume begins to rise at 6:00 and peaks between 9:00 and 18:00.

As shown in Fig. 10, among the transfer modes for passengers arriving at Hangzhou East Railway Station, the most popular is the metro. On average, 77% of arriving passengers use the metro (ranging from 67 to 83%, with higher arrival volumes corresponding to higher metro usage rates). Additionally, 6% of passengers choose to take buses, 9% opt for taxis, and 8% leave via private vehicles.

As shown in Fig. 11, although the metro is the most popular transportation mode, a significant proportion of transfer OD pairs fall outside the metro service area (i.e., the destination point is more than 1 km away from the nearest metro station). This group accounts for 60%–70% of passengers (holiday < weekend < weekday), meaning they cannot reach their destinations directly via the metro or by transferring within the metro system. Most passengers make a "leave first, decide later" choice, which is not their original intention.

Taxi queue lengths follow similar patterns. During 7:30–9:30 and 16:30–18:30, there are more people queuing for taxis. Additionally, after the metro service stops

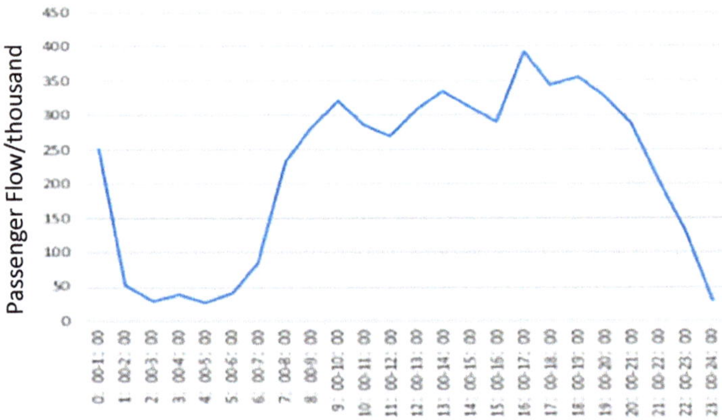

Fig. 9 Time segmentation of users arriving at and departing from Hangzhou East Station (Arriving on the Left/Departing on the Right)

Fig. 10 Transfer modes of passengers arriving at Hangzhou East Railway Station

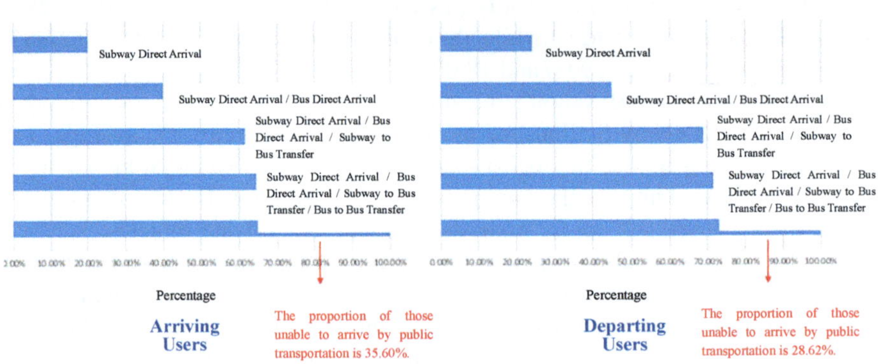

Fig. 11 Public transportation modes of arriving and departing users

operating, taxi queue numbers experience a slight increase. When passenger flow is high, the proportion of travelers choosing taxis decreases due to the longer waiting times for taxis. It is reported that the taxi queue at Hangzhou East Railway Station stretches up to 50 m, and there are often large numbers of people in line (Fig. 12). During periods of long taxi waiting times (10:00–20:00), operating customized buses could attract some of the passengers who would otherwise take taxis (Fig. 13).

Further Application: mobile signaling data can be used to classify users of various transportation modes and identify their dynamic activity trajectories. By treating all transportation modes as a unified system, this data can help integrate travel resources. For example, designing intermodal transit routes and adjusting bus

Fig. 12 Number of People Queuing for Taxis

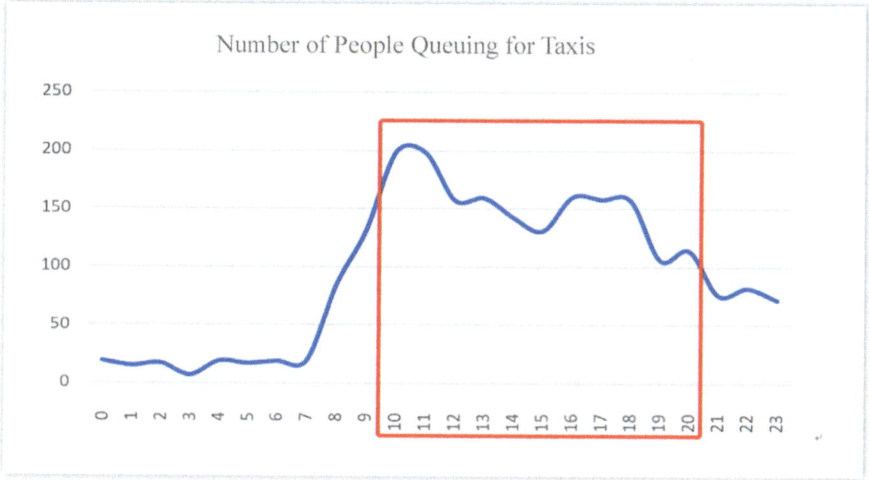

Fig. 13 Time segmentation of users at Hangzhou East Railway Station (Arriving on the Left / Departing on the Right)

stop locations can optimize and enhance the ability to manage public transportation. This would not only assist transportation management departments in accurately understanding rail transit line information and planning reasonable rail transit routes but also contribute to constructing a high-quality travel system. Such measures are significant for improving residents' quality of life and their satisfaction with urban transportation.

3.3 Evaluating Policy Implementation Effects

Similarly, mobile signaling data played a crucial role during the COVID-19 pandemic in observing the movement of people returning to their hometowns and assessing the risk of epidemic spread. In early January 2021, sporadic COVID-19 outbreaks were reported in multiple regions. Against this backdrop, on January 25, the General Office of the Central Committee of the Communist Party of China and the General Office of the State Council issued the Notice on Ensuring Services and Safeguards for People Staying in Place During the Spring Festival, Subsequently, various regions issued appeals to avoid unnecessary travel during the Spring Festival and introduced policies to encourage workers to celebrate the festival at their places of employment. For instance, Zhejiang, Fujian, and Guangzhou provided Spring Festival red envelopes; Shanghai, Hebei, Henan, Hunan, and Chongqing encouraged migrant workers to avoid unnecessary travel home, while local residents were urged to avoid unnecessary travel outside their provinces; Enterprises and institutions staggered leave schedules and implemented flexible vacation policies. Mobile signaling data during this period helped evaluate the effectiveness of these policies, identify the travel behavior of various groups, and analyze the heterogeneous impacts of the policies on return-home behavior.

Figures 14(a)–(d) present the absolute return-home proportions of different groups during the 2021 Spring Festival. Analysis of the return-home behaviors revealed that the return-home proportion in 2021 was only 30%–40% of the level in 2019. Groups with higher return-home proportions included high-income earners, low-education individuals, those aged 18–38, and males. Under the influence of the pandemic, the 2021 return-home proportion decreased compared to both 2019 and 2020. Groups such as middle-aged and young individuals, higher-income earners, and women saw a more significant decline in return-home behavior in 2021 due to the pandemic. Policies encouraging people to celebrate the Spring Festival in place generally promoted "staying in place" among all demographic groups.

Firstly, in terms of the spatial distribution of the relative proportions of return-home numbers during the Qingming Festival and Spring Festival in 2021 compared to 2019, the return-home population during the 2021 Spring Festival was only 45% of the same period in 2019. However, the number of people returning home during Qingming reached 1.76 times that of 2019, indicating a substitution effect between the two periods. Secondly, in 2021, individuals whose hometowns were farther away were more likely to choose to return home during Qingming. Figure 15 shows that

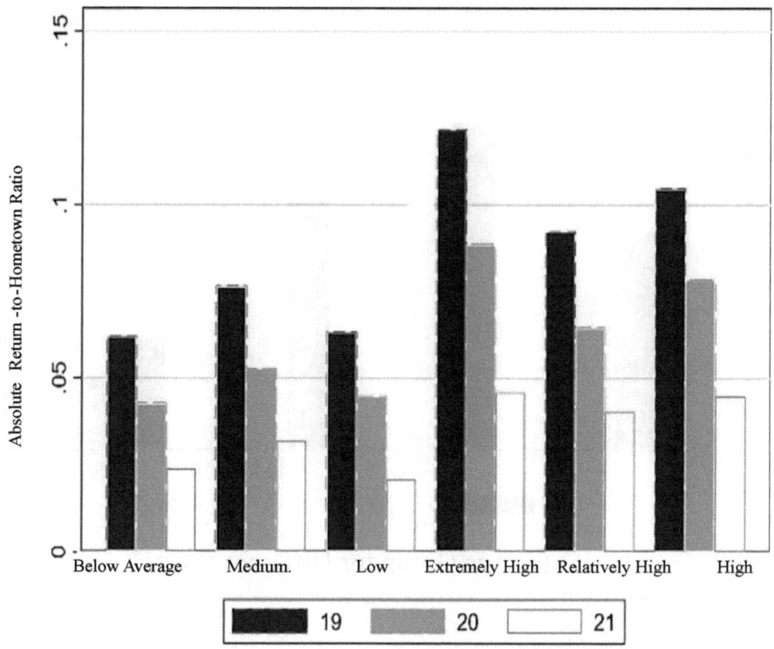

(a) Wealth Levels by Different Years

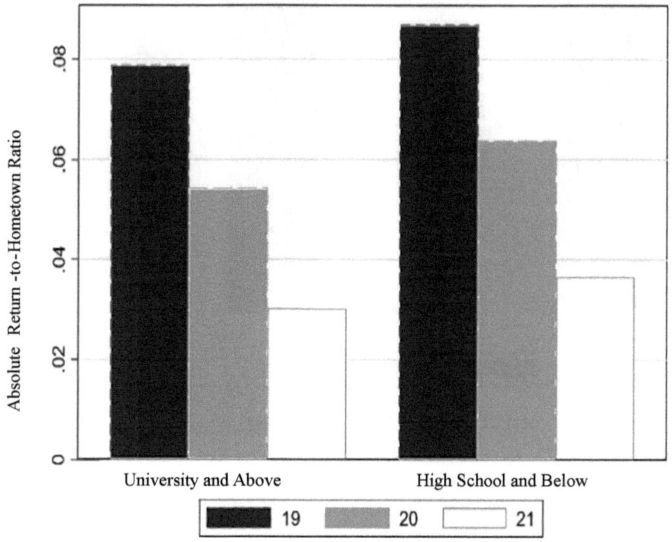

(b) Wealth Levels by Different Years

Fig. 14 Absolute return-to-hometown ratios of different groups during the 2021 Spring Festival

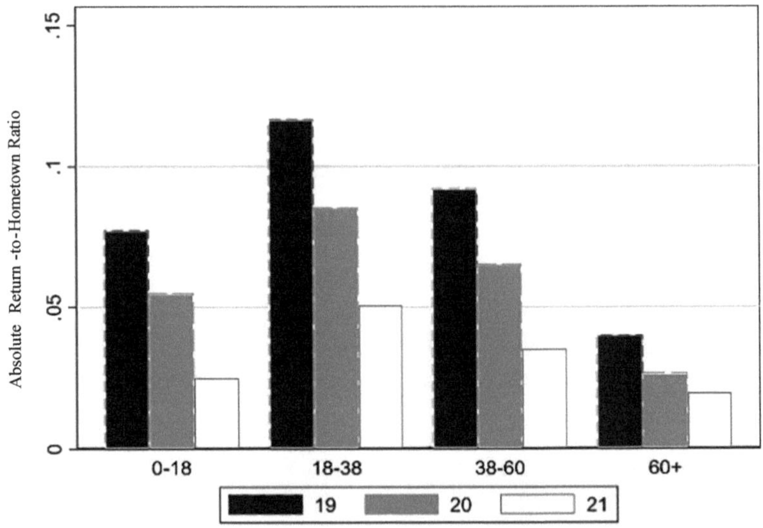

(c) Age Group for Different Years

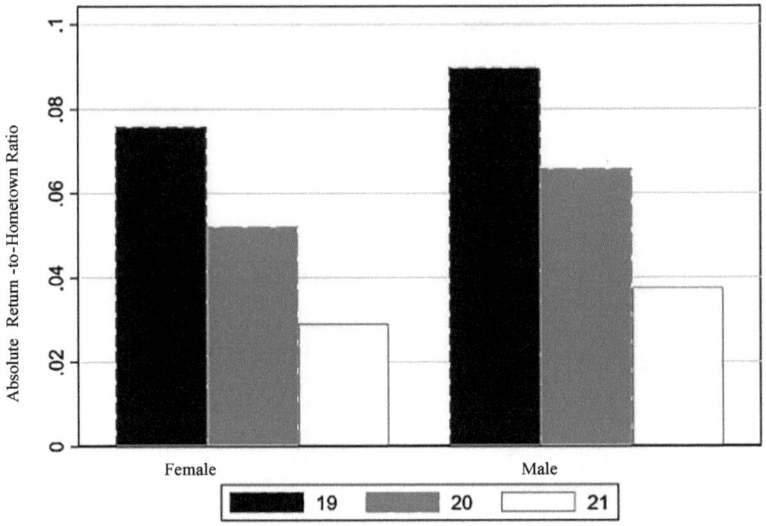

(d) Gender for Different Years

Fig. 14 (continued)

policies encouraging people to celebrate the Spring Festival in place promote deferred return-home behavior. Among these policies, cash-based incentives had a significant impact, while other types of policies were less effective.

Further Application: With the rapid development of computer technology and the continuous deepening of digital reform, the generation and rapid iteration of

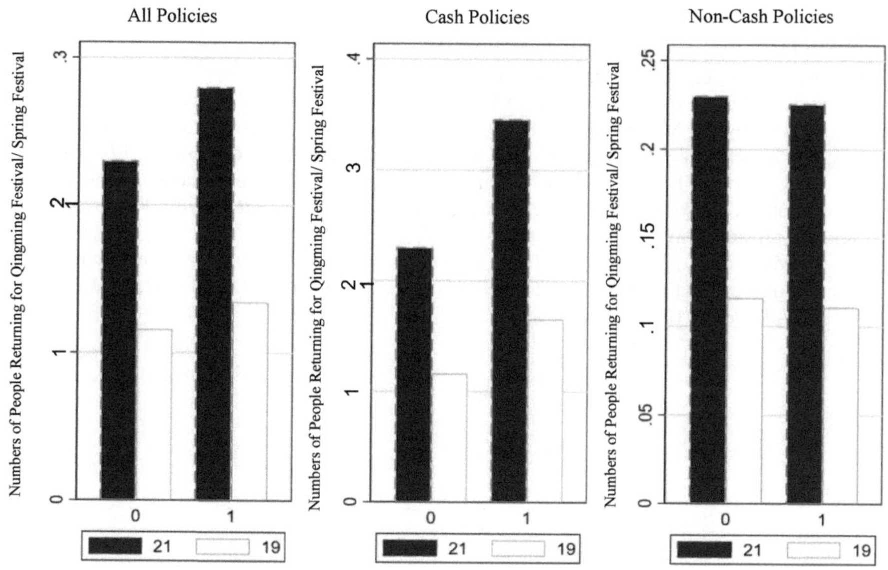

Fig. 15 Relative return-to-hometown ratios for the Spring Festival and Qingming Festival in Zhejiang Province: Based on Workplace

massive data have made it possible to transform traditional governance methods and extend their scope. For example, in emergency management scenarios, innovative governance tools based on big data (such as health codes) have emerged, weakening the need for multi-layered approval processes for traditional regulations. Similarly, the location information provided by mobile signaling data enables the dynamic management of inter-regional population flows. This not only aids in controlling social risks but also facilitates the recovery and development of production and daily life. Additionally, it allows for relatively timely assessments of policy effectiveness and the ability to adjust and improve policy directions promptly.

References

Costa, M. (2010). Interpersonal distances in group walking. *Journal of Nonverbal Behavior, 34*(1), 15–26. https://doi.org/10.1007/s10919-009-0077-y

Moussaïd, M., Perozo, N., Garnier, S., et al. (2010). The walking behaviour of pedestrian social groups and its impact on crowd dynamics. *PLoS ONE, 5*(4). Retrieved March 28, 2021, from https://journals.plos.org/plosone/article?id=10.1371/journal.pone.0010047

Rahman, S., & Smith, D. K. (2000). Use of location-allocation models in health service development planning in developing nations. *European Journal of Operational Research*, 437–452. https://doi.org/10.1016/S0377-2217(99)00289-1

Zhang, Y., Fu, W., Chao, H., Mi, Z., & Kong, H. (2024a). A comparative analysis of the potential of carbon emission reductions from shared micro-mobility. *Sustainable Energy Technologies and Assessments, 72*, Article 104088.

Zhang, Y., Kong, H., & Lin, D. (2023). Shared micromobility and future cities. *Transactions in Urban Data, Science, and Technology, 2*(4), 155–158.

Zhang, Y., Lin, D., & Mi, Z. (2019). Electric fence planning for dockless bike-sharing services. *Journal of Cleaner Production, 206*, 383–393. https://doi.org/10.1016/j.jclepro.2018.09.215

Zhang, Y., Manley, E., Martens, K., & Batty, M. (2024b). A metro smart card data-based analysis of group travel behaviour in Shanghai, China. *Journal of Transport Geography, 114*, 103764.

Zhang, Y., & Martens, K. (2018). An explorative analysis of group travel behaviour patterns in the public transit context. In *Transportation Research Board 97th Annual Meeting*. Retrieved March 30, 2021, from https://trid.trb.org/view/1495341

Zhang, Y., Martens, K., & Long, Y. (2018). Revealing group travel behavior patterns with public transit smart card data. *Travel Behavior and Society, 10*(October 2017), 42–52. https://doi.org/10.1016/j.tbs.2017.10.001

Chapter 6
The Future of Digital Urban Governance

1 Urban Governance in the Age of Artificial Intelligence

1.1 From Digital Governance to Smart Governance

In the future, artificial intelligence (AI) technologies will penetrate broader and deeper into all aspects of urban governance. Specifically, urban governance will increasingly rely on a series of AI technologies to achieve more refined management. Future cities will not only think intelligently but will also act more proactively and wisely. This will represent a significant shift from the "Internet of Everything" to the "Intelligence of Everything" and from perceptual intelligence to cognitive intelligence.

1.1.1 From Digitization to Smartification

Currently, the technologies widely applied in cities remain primarily information technologies. Urban governance heavily depends on the internet, mobile internet, and the Internet of Things (IoT). These technologies enable extensive connections among computer systems, mobile communication devices, and other urban facilities (e.g., buildings, infrastructure), as well as large-scale data collection, analysis, and processing through information collection terminals. This allows the operational patterns of cities to be clearly perceived and understood, enabling systems to accurately provide warnings, commands, and governance responses to urban matters. For example, the "Internet + Grid Management" governance model mentioned in Chap. 1 is a prominent case. In this model, governance focuses on grassroots issues, relying on grid officers to solve the "last mile" and "last step" problems. Moreover, the fine-grained grid governance model lowers the threshold for citizen participation, forming a robust supervision and feedback mechanism and fostering interactive multi-stakeholder governance (Zhang, 2018).

W. Zhang and Y. Zhang, *Digital Urban Governance in China*, Qizhen Humanities and Social Sciences Library, https://doi.org/10.1007/978-981-96-6325-5_6

As technology evolves, urban governance will transition from digitalization to intelligence and ultimately to smartification. From a technological perspective, smart governance mainly relies on AI technologies. While the core objectives of internet and IoT technologies are connection and perception—providing comprehensive and accurate information to support human decision-making—AI's ultimate goal is to liberate humans from complex urban tasks and overcome limitations posed by human resources. The "Hangzhou City Brain" case introduced in Chap. 4 exemplifies the application of AI in assisting urban governance. This type of intelligent urban operation center represents an early stage of smartification, achieving basic computer-driven decision management in traffic governance. Beyond establishing an integrated city brain with holistic thinking capabilities, smart urban governance emphasizes actionability (Gao & Liu, 2019). Actionability refers to the ability to execute the results of self-learning and deep learning through algorithms. Currently, urban actionability relies primarily on human-operated facilities and lacks independent action capabilities. Moreover, each node within a digital urban system network is responsible for data collection, analysis, and transmission but is constrained by the overall network's efficiency.

In the realm of smart urban governance, AI chips are key to creating independent action units in smart cities, breaking through the limitations of traditional intelligent terminal workflows (Gao & Liu, 2019). Traditional intelligent terminal workflows typically rely on a centralized AI server (e.g., the City Brain), which integrates various algorithms to analyze urban data and user demands collected by terminals (sensors, cameras, mobile devices). After processing, the information is transmitted back to the terminals and users. However, if any part of this process encounters a problem or transmission is disrupted, the entire intelligent urban center is affected. AI chips allow each terminal to conduct personalized self-learning and automatically make decisions based on user information without needing input from or output to a central "brain" server. As urban governance transitions into the smartification stage, each terminal in a city will become an independently capable unit with autonomous learning and action abilities.

1.1.2 From "Internet of Everything" to "Intelligence of Everything"

IoT technology has made the "Internet of Everything" a reality in cities, featuring applications in smart pipelines, smart firefighting, smart parks, and smart agriculture. A key characteristic of IoT development is the advent of the era of big data, which cannot rely solely on human efforts for tasks such as data filtering, analysis, and mining. Thus, the integration of artificial intelligence (AI) and IoT has become an inevitable trend. AI, serving as the top-level data processing center for IoT, is driving the world toward the "Intelligence of Everything".

In recent years, IoT has become a focal point in advancing urban smartification and intelligence. Various sensors spread across cities have made information collection both simple and efficient. IoT, as "the Internet of connected things", is an integral part of next-generation information technology. Its core foundation remains the internet, but its user base has expanded to include connections between any objects for information exchange and communication (Hengxing IoT, 2021). Initially, IoT was merely a tool for connecting and collecting information between objects. With the arrival of the "Internet of Everything" era, IoT's scope has expanded from IoT (Internet of Things) to IoE (Internet of Everything) (Yang et al., 2016). By 2020, the number of IoT connections surpassed non-IoT connections for the first time, signifying that the two primary goals of IoT technology development—"more thorough perception" and "more comprehensive interconnection"—had been largely achieved. The current focus has shifted to "deeper intelligence", marking the transition from the "Internet of Everything" to the "Intelligence of Everything". The integration of AI and IoT has accelerated this shift, giving rise to a new term: Artificial Intelligence of Things (AIoT). AIoT combines AI and IoT technologies, referring to the generation and collection of massive amounts of data by IoT devices, which are then uploaded and stored in the cloud for in-depth analysis by various AI technologies. This enables the digitization and intelligent interconnection of all things. The smart devices created through AI and IoT integration can make precise and wise decisions without human intervention.

AIoT comprises a four-layer model, listed from top to bottom: perception and recognition layer, network transmission layer, information processing layer, comprehensive application layer. Simply put, AIoT is based on the information collected by the perception and recognition layer, transmitted through the network transmission layer, autonomously analyzed and utilized by the information processing layer, and then fed back to terminal devices with specific instructions for execution. Firstly, perception and recognition layer employs information sensing devices such as RFID, infrared sensors, GPS, laser scanners, and embedded systems to connect any object to the internet. These devices gather object-related information through identification, positioning, tracking, and monitoring. AI enhances sensor capabilities, improving data usability. Secondly, network transmission layer uses various networks, including the internet, wireless broadband, low-speed wireless networks, and mobile communication networks, to transmit large volumes of information. Acting as a bridge, it transmits data collected from the perception layer to the information processing layer for analysis and sends the resulting commands back to the terminals. Thirdly, information processing layer leverages AI's autonomy to address the challenges of data storage, classification, and security. Algorithms analyze data to generate actionable instructions. AI applications at this layer improve human work efficiency, optimize workflows, and enable deeper data analysis to enhance decision-making precision, innovation, and systematization. Fourthly, comprehensive application layer focuses on implementing instructions generated by the information processing layer. It encompasses IoT terminals designed by various industries, enterprises, or domains to meet their specific needs.

The integration of IoT and AI is the inevitable result of both technologies' development. IoT requires AI to unlock its immense potential and expand its application boundaries. Similarly, AI needs IoT as a platform to facilitate implementation. Thus, AI and IoT are complementary entities whose combined force will strongly drive cities toward the "Intelligence of Everything".

1.1.3 From Perceptual Intelligence to Cognitive Intelligence

Currently, the widespread application of various technologies in cities has enabled many of them to possess basic capabilities such as "hearing, seeing, and perceiving". To accelerate the modernization of urban governance, it has become a prevailing trend to enhance the intelligence of urban services and policy decision-making, expand the scope of artificial intelligence (AI) applications, and drive AI technology toward the higher level of "strong AI".

AI technology can be categorized into three levels: computational intelligence, perceptual intelligence, and cognitive intelligence. As the names suggest, computational intelligence refers to rapid calculation, memory, and storage capabilities; perceptual intelligence involves sensory abilities such as hearing, vision, and touch; cognitive intelligence, however, is far more complex, encompassing abilities like analysis, reasoning, understanding, and judgment. With the continuous development of computational power and advances in storage methods, computational intelligence has already been realized. In terms of perceptual intelligence, current AI technology has reached or even surpassed human levels. For example, facial recognition and speech recognition are two major applications representing perceptual intelligence. However, cognitive intelligence remains out of reach. Take the currently popular facial recognition technology as an example: this technology merely performs matching tasks and lacks the subsequent processes of information processing, understanding, and reasoning that occur in the human brain. Therefore, this technology falls within the realm of perceptual intelligence rather than cognitive intelligence. At present, both academia and industry generally agree that AI, having achieved computational and perceptual intelligence, is now striving toward the goal of cognitive intelligence. In January 2020, Alibaba DAMO Academy released the "2020 Top 10 Technology Trends", identifying "the evolution of AI from perceptual intelligence to cognitive intelligence" as one of these trends. Specifically, "Cognitive intelligence will draw inspiration from cognitive psychology, neuroscience, and human societal history, combined with cross-domain knowledge graphs, causal reasoning, and continuous learning technologies, to establish effective mechanisms for acquiring and expressing knowledge stably. This will enable machines to understand and apply knowledge, achieving a critical breakthrough from perceptual intelligence to cognitive intelligence" (Alibaba DAMO Academy, 2020). Simply put, cognitive intelligence involves three stages: understanding, analysis, and decision-making. During the understanding stage, knowledge graphs are constructed based on the knowledge base from the perceptual intelligence phase. In the analysis stage, hidden and explicit relationships between data are identified using the knowledge graph. Finally, in the decision-making stage, executable solutions are provided (Su Lun AI Channel, 2020). Achieving cognitive intelligence is an enormous challenge because it entails enabling AI to build new neural network structures that mimic human thinking, with the capacity to acquire knowledge, reason, and draw associations like the human brain.

If computational intelligence and perceptual intelligence rely primarily on algorithms and data, then achieving cognitive intelligence hinges on enabling AI to acquire human knowledge and common sense. Developing knowledge graph technology has become a key breakthrough in this process. A knowledge graph refers to the techniques used for constructing and applying knowledge graphs, representing one of the three major paradigms of AI research—specifically, the symbolic AI paradigm (Li et al., 2019). It was first introduced by Google in 2012 and successfully applied to search engines. This technology integrates cognitive computing, knowledge representation and reasoning, information retrieval and extraction, natural language processing, semantic web, data mining, and machine learning. At its core, it involves the efficient processing and visualization of multimodal, multi-source heterogeneous data and complex multidimensional relationships, transforming relational attributes that are hard to represent with mathematical models into a data network connected by relationships (AI Chip World, 2021). The knowledge graph describes concepts, entities, and their interrelationships in the objective world in a structured form, organizing and managing the vast information on the internet in a manner closer to human cognition (Li et al., 2017). Within a knowledge graph, attention must be paid not only to the richness of knowledge points but also to their interconnections, as these associations enable AI to form connections and draw inferences. The denser and more complex these connections are, the "smarter" AI becomes, and the higher its level of cognitive intelligence. Together with big data and deep learning, knowledge graphs have become one of the core drivers of AI development.

In the context of smart cities, knowledge graphs are the most effective and direct method for processing big data due to their data structures closely resembling the real world and aligning with human cognitive patterns. By leveraging knowledge engineering, semantic knowledge can be added to multi-source heterogeneous data, allowing it to generate "intelligence" and enabling a leap from data to information and eventually to knowledge. Knowledge graphs are also critical enablers for smart city development, transforming human knowledge, experience, logic, and methodologies into productivity. By linking and converting relationships between concepts, knowledge graphs facilitate the organization of business rules, visualization of process controls, and exploration of deeper relationships among phenomena. Moreover, knowledge graphs provide robust support for applications in urban living, government operations, and policy-making. Using an open standard unified framework, they employ a homogenous classification method and vocabulary to describe data and common terminology shared across all sectors in a city. This facilitates the semantic linkage of multi-source data from various domains, such as daily life, production, and governance, on a semantic graph. The result is a dynamic knowledge base that visually reveals related points in relationships, enabling easy organization, management, and utilization of knowledge by humans (Feng & Qi, 2021; Jin, 2019).

1.2 *Future Prospects of Artificial Intelligence in Urban Governance*

1.2.1 Smart Transportation

The establishment of urban smart transportation systems based on artificial intelligence technology has become a crucial trend in urban traffic management. Intelligent traffic signal control and autonomous driving, as two essential components of smart transportation systems, contribute significantly to improving urban road efficiency.

One of the key reasons for slow road traffic flow and low road resource utilization is the unreasonable setup of traffic signals. The application of artificial intelligence technology facilitates the construction of adaptive traffic signal control systems. These systems can monitor and predict road information, automatically calculate traffic flow, and generate automated analysis and decision-making processes. This allows for scientific optimization of signal timings, thereby reducing vehicle wait times at intersections. For example, in July 2020, Suzhou High-Tech Zone in Jiangsu Province implemented the province's first adaptive traffic signal control system. The system, equipped with pedestrian density and dynamic thermal imaging detectors, as well as pedestrian crossing warning columns, significantly enhanced pedestrian crossing safety and traffic efficiency while enabling all-day automatic signal timing control functionality (Suzhou Daily, 2020).

Promoting autonomous driving can also significantly enhance traffic efficiency. Autonomous vehicles leverage artificial intelligence, cloud computing, and robotics to plan routes scientifically, identify the optimal driving paths, and intelligently detect and avoid obstacles, thereby improving traffic flow in all dimensions. Autonomous driving technology consists of three key components: perception, decision-making, and execution. Perception layer utilizes sensors such as cameras and radars mounted on the vehicle to collect and process environmental data, enabling the detection of surroundings. Upon receiving images and data from cameras, the autonomous driving system uses deep learning and other technologies for target detection, accurately segmenting and classifying objects such as the environment, other vehicles, pedestrians, and traffic infrastructure. Built on deep learning and computer vision algorithms, decision layer processes the sensor data in real-time, automatically recognizing road conditions and predicting the behavior of vehicles, pedestrians, and other road users. Based on these predictions, the decision layer makes behavioral decisions for the vehicle, issuing control instructions. The decision layer also handles multi-source data processing, path planning, behavior planning, speed and acceleration control, and trajectory planning. Execution layer controls the vehicle based on the decisions made by the system. It connects the vehicle's various control systems to the decision-making system via a bus network. These systems then accurately execute driving actions according to the control commands from the decision-making system, achieving autonomous driving (Snowball, 2020). Currently, autonomous driving faces several challenges, including ethical dilemmas and safety liabilities.

However, California has already legalized the use of autonomous vehicles on public roads, demonstrating the feasibility of this technology (China News Network, 2012).

1.2.2 Smart Education

The expansion of personalized learning opportunities is one of the key indicators of improved educational quality in the new era, and the application of artificial intelligence (AI) in education provides students with more opportunities for personalized learning. The new generation of children and adolescents is growing up alongside the rapid development of technology. Traditional learning models, which are uniform in pace and fixed in time and location, can no longer meet their higher learning demands. AI technology makes it possible to fulfill these demands. Firstly, through data mining and deep learning technologies, big data on students' learning behaviors can be analyzed to offer targeted and appropriate learning paths and resources based on individual student needs and circumstances. Secondly, AI-related technologies can also create immersive virtual learning environments, enabling students to study at any time and place. Thirdly, AI can track and analyze changes in students' cognitive levels and emotional states, and respond intelligently to ensure that students engage in learning with a positive attitude. For example, intelligent teaching systems simulate students' cognitive and emotional states, matching learning activities to their cognitive needs and emotional states to ensure deep engagement during the learning process (Liu et al., 2018).

With AI technology, teaching quality and learning efficiency can be significantly improved even in scenarios where teachers and students are separated by time and space, thereby ensuring the effective implementation of remote education. Typically, such applications rely on external devices such as computers, wearable devices, and cameras. By analyzing data from these devices, AI can continuously track the behaviors of students and teachers, analyze behaviors in specific scenarios, and make targeted decisions. These decisions are then supported by adaptive learning systems, which push matched learning content, teaching experts, and resources to different users.

Moreover, through smart teaching systems, intelligent question-answering systems, smart learning games, intelligent administrative systems, and decision-support systems, AI can play a pivotal role in education. It can replace teachers and other educational staff in handling repetitive, tedious, and low-efficiency tasks, freeing up more time and energy for teachers to think about and design more innovative educational processes and methods (Zhang et al., 2017).

1.2.3 Smart E-Government

The application of artificial intelligence (AI) in government services is becoming increasingly widespread. Lin (2018) pointed out that the current use of AI in government services primarily includes identity verification, intelligent customer service, smart search, situational awareness, and intelligent robots. Lin also suggested that AI could eventually support decision-making in government services based on big data. Similarly, Qiu (2017) studied the applications and challenges of AI technology in government services and proposed that its applications mainly include verification tools, content recommendations, online customer service, and information processing. From a goal-oriented perspective, AI can enhance the quality and efficiency of government services through three main pathways:

Firstly, it can enhance government service and management efficiency. Traditional administrative approval processes often involve numerous approval steps, complex workflows, and difficulties in inter-departmental coordination. On the one hand, AI technology can accurately analyze and streamline the various stages of government service processing, identify redundant steps, and simplify them, thereby improving service efficiency. On the other hand, the automation capabilities of AI allow it to process certificates, documents, and official papers, as well as transmit a wide range of service data, at speeds far exceeding human capabilities. This helps to eliminate data silos caused by departmental fragmentation in traditional government services, enabling efficient cross-departmental and cross-regional collaboration in e-government services (Chen et al., 2018).

Secondly, it can improve the scientific and accurate nature of government decision-making. Decision-making based solely on experience has gradually revealed its shortcomings. With the extensive application of AI technology, government discretion is significantly reduced, thereby promoting social equity. Using various AI algorithms, government departments can conduct deep analyses of data to uncover public needs, enabling more precise and efficient public services. For business-related services, governments can utilize their extensive databases of corporate and industrial information. With the assistance of AI technology, they can generate economic analysis reports containing data on industry trends, market activity, and business vitality. These reports not only provide a basis for economic governance decisions but also enable the government to deliver personalized policies to businesses, achieving high-quality proactive services (Ai & Wang, 2015).

Thirdly, it can enhance the efficiency of government-citizen interaction. Traditional government service processes are often constrained by time and distance, meaning the public can only interact with government staff in person during limited hours. This frequently leads to unresolved issues and inefficient communication. AI's human-like interactive capabilities, coupled with its precise understanding of public information, make it a valuable tool for improving government-citizen interactions. Such technologies can also operate 24/7, replacing human efforts to provide a superior interactive experience. Furthermore, AI can collect and analyze public feedback on government service processes, staff attitudes, and other aspects, allowing governments to pinpoint areas requiring improvement and make necessary adjustments (Chen et al., 2018; Li, 2008).

1.2.4 Smart Healthcare

It is estimated that the total market value of artificial intelligence (AI) applications will reach $127 billion by 2025, with the healthcare industry accounting for one-fifth of the overall market size (Internet Medical and Healthcare Industry Alliance, 2018). The application of AI in the healthcare industry primarily benefits from the continuous expansion of medical databases and the increasing capability of machine learning to analyze medical data.

AI has already shown promising results in health management. By analyzing genetic data, metabolic data, and phenotypic data, AI technology can provide targeted health recommendations for residents, covering areas such as diet and lifestyle (Dong & Wang, 2017). This helps mitigate health risks in daily life and prevent diseases. When necessary, AI can also activate risk warning functions.

As the aging population in China continues to grow, the issue of inadequate medical resources becomes more acute, highlighting the crucial role of AI in elderly health management. Intelligent devices such as sensors, wearable devices, and smart nursing beds are increasingly being used in this field. The development of "AI + Healthcare" has also led to the emergence of new industries and models, such as health service robots and companion robots. For example, the Husky robot from Tianjin is being promoted nationwide in hospitals and clinics. In Hangzhou, robots are widely used in welfare centers and elderly care institutions, with over 70 elderly care facilities and 40 care centers introducing the "A-Tie" robot, which offers four primary functions: health monitoring, health consultancy, emergency alerts, and companionship (Du, 2017; Wang, 2017).

AI can also improve the quality of healthcare services through applications such as smart medical imaging diagnostics and surgical robots. Smart medical imaging diagnostics involve using computers to analyze medical images through deep learning, enabling tasks such as image classification, target detection, image segmentation, and retrieval. These tools assist doctors in diagnosis, treatment, and other clinical work (Zhao et al., 2018). One of the most important applications of smart medical imaging diagnostics is cancer detection. In 2017, scientists from Google, Google Brain, and Verily developed an AI product capable of diagnosing breast cancer with an accuracy rate close to 90%, significantly outperforming the average accuracy rate of around 70% achieved by pathologists (Gao & Lü, 2017). AI is also increasingly being applied in treatment, with one of the most critical applications being surgical robots. Under current technological capabilities, surgical robots can effectively assist doctors during operations, particularly in minimally invasive surgeries, and are continuously advancing toward higher stages of development. For instance, in May 2016, a research team from the Children's National Health System in the United States developed the Smart Tissue Autonomous Robot (STAR) to enhance surgical efficiency and safety. STAR is the world's first autonomous surgical robot capable of handling soft tissues. This invention signifies that, with technological advancements, robots may eventually take over the work of surgeons, providing surgical and other medical services for humans in the future (Gao & Lü, 2017).

2 Opportunities Brought by Data Openness to Urban Governance

In the era of big data, cities generate massive amounts of data every day. As a critical production factor and core resource, data openness, particularly government data openness, has become a crucial lever for enhancing urban governance efficiency and achieving high-quality economic development.

With the support of emerging technologies, the construction of digital governments has been advancing steadily, accumulating vast amounts of data closely related to citizens' lives, business operations, and societal functions. On August 31, 2015, the State Council issued the Action Plan for Promoting Big Data Development, which explicitly called for "accelerating government data openness and sharing, driving industrial innovation and development, fostering emerging industries, and supporting economic transformation". On October 29, 2020, the Central Committee of the Communist Party of China proposed in its Recommendations for Formulating the 14th Five-Year Plan for National Economic and Social Development and Long-Range Objectives for 2035 to "accelerate digital development, establish fundamental institutions and standards for data resource property rights, trading, cross-border transmission, and security protection, promote the development and utilization of data resources, expand the orderly openness of basic public information data, and build a unified national platform for data sharing and openness". At present, promoting government data openness has become an unstoppable trend. Many cities across the country have established public data platforms as vehicles for government data openness. For instance, Zhejiang Province passed the Regulations on Public Data of Zhejiang Province during the Sixth Session of the 13th People's Congress on January 21, 2022, which took effect on March 1. It is the nation's first local regulation in the public data domain and aims to fully realize the value of public data as a new production factor, offering a Zhejiang model for modernizing governance capabilities.

However, data openness still faces challenges such as an incomplete legal framework, severe data fragmentation, and a shortage of data talent. It is therefore urgent to address these issues through measures such as improving the legal system for data openness, establishing secure and scientific data storage systems, and promoting the market-based allocation of data resources. These steps are essential for strengthening the foundation of the data resource system and fully unleashing the potential of data dividends.

2.1 Significance of Data Openness

2.1.1 Enhancing Government Governance Capabilities

Cross-departmental data openness and sharing are essential for enhancing governance capabilities in the new era, as they play a significant role in improving departmental collaboration and elevating the government's ability to serve its citizens comprehensively. Only when government data is fully interconnected internally can orderly and effective openness to society at large be realized.

Historically, technological and institutional constraints have resulted in significant fragmentation of government data storage, creating numerous "data silos" where information between departments could not be interconnected, shared, or integrated. This not only increased the difficulty of interdepartmental communication but also made administrative procedures unnecessarily rigid and complex for citizens. In the age of big data, the establishment of unified government data platforms and institutional frameworks for cross-departmental data sharing has enabled the integration of data across different government sectors. This resolves the problem of "data silos", as data scattered across departments is now centrally aggregated and stored. Consequently, the efficiency of cross-departmental processes has been significantly improved. Furthermore, based on unified algorithm models, cross-departmental and cross-sectoral data can undergo intelligent and refined analysis, providing scientific policy support for public services, economic management, and environmental governance.

2.1.2 Protecting Citizen Rights

From a legal perspective, the right to know is widely recognized as the foundation or a key component of government information openness (Chen, 2016; Zhang, 2020). Article 1 of the Regulations on the Disclosure of Government Information of the People's Republic of China explicitly states that government information serves the production, living, and socio-economic activities of the people. If data and information are concentrated solely in the hands of the government and cannot be effectively accessed or utilized by citizens, it hinders the formation of their thoughts and opinions, thus compromising their freedom of expression (Zhang et al., 2002). Therefore, to ensure data is "sourced from the people and used for the people", the government should proactively make data available to all citizens for free, without requiring authorization, and in a non-discriminatory manner. Moreover, the data should be provided in multiple formats to cater to the needs of different types of users (Fu & Zheng, 2013).

Government data openness can drive transparency in government operations, protect citizens' rights, and promote civil society development. First, open data is a key measure to empower the public. With access to open data, citizens can better understand government policies, enabling them to present demands and opinions more effectively, which helps government departments refine their policies accordingly. Second, individuals and organizations familiar with data science can analyze and interpret open government data, uncover deeper insights, and provide comprehensive perspectives to assist government decision-making. Third, data openness supports public oversight. Transparent and efficient administrative processes can significantly improve public satisfaction and trust in the government. For example, Shanghai initially began publishing government data by focusing on highly scrutinized departments with significant public interest, such as those managing fiscal budgets, administrative fees, government procurement, major construction projects, and urban planning (*Liberation Daily*, 2016). This initiative safeguarded citizens' rights to know and supervise, facilitating the implementation of relevant government policies.

2.1.3 Driving Social Integration and Innovation

When government data remains inaccessible, businesses and individuals often miss opportunities for innovation due to insufficient data resources. Thus, open data is vital for fostering social integration and innovation.

Government data openness promotes the development of new technologies and business models within cities. Open data provides a strong foundation and conducive environment for derivative innovation in fields such as big data, cloud computing, mobile internet, the Internet of Things (IoT), and smart logistics. It also facilitates the innovation of emerging industries and sectors. With access to open government data, businesses can acquire data more quickly and conveniently, utilize it to create additional commercial value, and enhance their innovation capabilities and core competitiveness. This, in turn, drives industry and sector innovation. For instance, the U.S. government's weather data openness gave rise to a robust weather risk management industry, which boasts an output value 60 times that of Europe and 146 times that of Asia (Tang et al., 2018).

For the public, government data openness reduces the costs and barriers to innovation, encourages value co-creation, and stimulates the innovation vitality of diverse stakeholders. This further enhances digital urban governance and drives innovation in public service provision. For example, Zhejiang Province organized a data openness innovation competition, encouraging businesses, social organizations, and individuals to develop innovative applications using publicly available government data. These applications addressed key public service areas such as transportation, health, education, ecology, and public safety. Additionally, the competition encouraged participants to explore the innovation potential of social data and foster its integration with government data, thereby achieving efficient utilization of multi-source big data.

2.2 *Future Prospects for Data Openness*

2.2.1 Establishing a Comprehensive Legal Framework for Data Openness

A robust and practical legal framework is essential to ensure the smooth implementation of government data openness and the orderly participation of citizens and enterprises in data development and utilization.

Firstly, a legal and regulatory system grounded in local data openness practices should be established. Currently, many cities in China, such as Beijing, Shanghai, Hangzhou, and Guiyang, have accumulated considerable practical experience in data openness, yielding numerous outstanding achievements. However, local legislative powers are limited, and practical exploration often struggles to address complex real-world issues. Therefore, local governments should seek greater legislative authority from the central government regarding data openness. Simultaneously, during rule-making, it is crucial to engage more legal and big data experts to propose rules or solutions tailored to actual conditions. For example, Zhejiang Province issued the Interim Measures for Public Data Openness and Security Management in June 2020, covering areas such as data openness, utilization, security, supervision, and legal responsibilities. Similarly, on November 29, 2021, the Shanghai Municipal People's Congress released the Shanghai Data Regulations, which include provisions on general principles, data rights protection, public data, data element markets, data resource development and application, data reform in the Pudong New Area, data cooperation in the Yangtze River Delta region, data security, legal responsibilities, and supplementary articles.

Secondly, data governance capacity should be reinforced by clearly defining data quality requirements in the legal framework. Data quality serves as a litmus test for a city's data governance capabilities and reflects the local government's governance proficiency. Effective data governance must begin with strict quality control at the source, with legal and regulatory provisions clearly specifying data quality standards. Currently, many cities face inconsistencies in data upload requirements, formats, and content across various departments and levels. Moreover, government data portals often lack comprehensive and complete metadata. Moving forward, a standardized framework for government data openness should be developed and implemented. This includes establishing mechanisms to ensure data collection quality and accelerating the creation of a standardized data openness framework for public institutions, including government departments and affiliated organizations (Song, 2018). For example, the United States has introduced a series of laws governing data openness, such as the Freedom of Information Act, Privacy Act, and Open Government Directive. Regarding technical standards, the U.S. has also established metadata governance standards and other technical regulations to ensure data openness, effectively standardizing data openness practices (Wang, 2021).

Thirdly, it needs to further clarify data ownership, especially citizens' data rights. In 2010, UK Prime Minister David Cameron first introduced the concept of the "right to data", asserting that in the information age, every citizen should enjoy data rights (Ma et al., 2016). From the perspective of China's current data rights framework, legislative proposals to gradually improve the ownership and rights associated with data were explicitly put forward during the First Session of the 13th National People's Congress in 2018. The process of establishing national legislation on data rights is accelerating, and local governments have also initiated relevant legislative practices. For example, on July 15, 2020, the Shenzhen Municipal Bureau of Justice released the Shenzhen Special Economic Zone Data Regulations (Draft for Comments), which was the first official document in China to propose that individuals have data rights. The innovative aspect of this draft lies in its explicit definition of data rights as a person's legal authority over specific data, including autonomous decision-making, control, processing, deriving benefits, and receiving compensation for damage to interests. It also proposed, for the first time at the policy level, that natural persons, legal entities, and unincorporated organizations enjoy data rights as stipulated by laws, regulations, and the Regulations (*Shenzhen Special Zone Daily*, 2020).

2.2.2 Establishing a Scientific Data Security Storage System

For the vast and multi-source data generated by cities, it is critical to prioritize data security. The future development of digital cities must establish a robust data security storage system that not only addresses traditional threats such as data destruction and loss but also prevents the leakage of citizens' private information and e-government data.

In terms of protecting citizens' data, Wenzhou City has provided a model for protecting personal data privacy. In October 2019, Wenzhou launched the "Personal Data Vault" application system on the provincial and municipal public data platforms. This system allows individuals to manage their own personal data while offering a range of data security services to protect personal data ownership and privacy. It also prevents city government departments from misusing public data in their official duties, thereby enhancing government credibility. Specifically, the system uses blockchain technology to create personal data cloud credentials. It returns data access rights, portability, correction, autonomous decision-making, awareness, and usage rights to data owners. When a government department requires access to a citizen's personal data, authorization from the citizen is mandatory via a blockchain-generated data identification number. Furthermore, all platform data access logs are recorded on the blockchain, ensuring traceability and immutability. Users can review their data access logs through the "Access Records" feature and file complaints with the Municipal Big Data Bureau if unauthorized access is detected (Zhejiang Provincial "Run Once at Most" Reform, 2020).

In terms of protecting urban data, to enhance the security of its data infrastructure, Wuxi City in Jiangsu Province signed a strategic cooperation agreement with the National Computer Network Emergency Response Technical Team in July 2020 to establish the Wuxi City Network Security Comprehensive Management Platform. The "Xidun" Network Security Comprehensive Management System platform, which is part of this initiative, is now operational. It provides uninterrupted security monitoring for 5,000 websites and 1,000 key information systems across government agencies and critical enterprises. The platform detects zombie websites, malware, and worm viruses across 10,000 city websites. Since its trial phase, it has identified and reported over 200 network security vulnerabilities, all of which have been promptly resolved. This project has established a comprehensive city-level network security protection system, safeguarding the development of "Digital Wuxi" and was included in the "2021 Jiangsu Province Excellent Cybersecurity Practice Cases" (China Jiangsu Net, 2021).

2.2.3 Promoting Market-Based Allocation of Data as a Factor of Production

In 2020, China issued the Opinions on Improving the Mechanisms for Market-Based Allocation of Factors of Production, which for the first time explicitly recognized data as a new type of production factor at the policy level. This document also pointed the way for advancing market-based reforms of data factors. Under policy guidance, governments and enterprises have actively participated in building the data factor market and exploring feasible paths for data value transformation.

The core of promoting market-based reform of data factors is to establish a sound market mechanism, promote the construction of data trading markets, and provide standardized channels for circulation and transactions. For example, cities such as Guiyang, Beijing, and Shanghai have built data exchanges to promote the compliant flow and trading of data. On November 25, 2021, the Shanghai Data Exchange was officially inaugurated. To address key challenges such as data ownership, pricing, trust, access, and supervision, the Shanghai Data Exchange introduced a series of innovative measures, many of which are the first of their kind nationwide. First is to establish the nation's first data trading ecosystem. This system covers data trading entities, compliance consulting, quality evaluation, asset appraisal, delivery, and other areas, fostering and regulating new entities while creating a prosperous data circulation and trading ecosystem. Second is to establish the nation's first regulatory system for data trading. With the principles of "no compliance, no listing" and "no scenarios, no trading", a series of institutional norms governing the entire data trading process were established. Thirdly is to build the nation's first fully digital intelligent data trading system. This ensures full-time listing, global trading, and full traceability of data transactions.

Fourth, introducing the nation's first data product registration certificate: by issuing registration certificates and transaction certificates for data products, a unique identifier is established for each data set, enabling registration, statistics, and audits.

And finally, publishing the nation's first data product manual: by transforming abstract data into tangible products through product manuals, data is made more comprehensible and actionable (CCTV News, 2021).

2.2.4 Encouraging Cross-Regional Cooperation in Data Openness

In recent years, cooperation in areas such as economic development, public services, and environmental protection between cities has deepened, with multi-center urban clusters and coordinated urban regional development models becoming increasingly prevalent. In the digital era, digital cities not only need to fully explore and extract the value of data within their jurisdiction but also explore regional cooperative data openness models, forming mechanisms and technical pathways for data sharing and exchange, thereby facilitating the high-quality coordinated development of digital cities.

Current mainstream trends include establishing intercity data cooperation channels and unified regional data-sharing platforms. One representative case is the Chengdu-Chongqing city cluster. In recent years, 16 cities in Sichuan and Chongqing have signed 78 big data cooperation agreements, including strategic frameworks such as the Sichuan-Chongqing Cyber Information Collaboration Framework Agreement and the Integrated Development Demonstration Zone for Industrial Internet Strategic Cooperation Agreement. These agreements aim to advance big data collaborative development, jointly build regional big data resource centers, and establish regional urban brains. Additionally, the provincial and municipal governments of the two regions have jointly established a "Chengdu-Chongqing Cross-Region Service Zone", utilizing cross-regional citizen data for collaborative services. By 2021, Chengdu and Chongqing had interconnected data from 144 departments and 5,460 types of urban data, co-established 210 public service items, and processed an average of 15,000 cases per day using cross-border, multi-field, cross-department data calls, thereby empowering collaborative development within the urban cluster (Chongqing Daily, 2021).

For intercity data cooperation channels, Jiangmen City in Guangdong Province has provided an integrated data interconnection solution with the Macau Special Administrative Region (SAR) and Hong Kong SAR. In April and September 2021, the "Jiangmen-Macau Cross-Border Administrative Service Zone" and the "Jiangmen-Hong Kong Cross-Border Administrative Service Zone" were launched. By leveraging blockchain, communication technology, and data-sharing techniques, the three cities established collaborative governance channels. Citizens in the three regions can use remote video services to handle about 400 administrative matters, including business registration, taxation, and social security services. This initiative deepened digital city integration among Jiangmen, Hong Kong, and Macau (Jiangmen Broadcasting and Television Station, 2021).

2.2.5 Promoting Data Openness to the Whole Society

According to Fudan University's China Local Government Data Openness Report (Second Half of 2020), as of October 2020, 6% of provincial administrative regions (excluding Hong Kong, Macao, and Taiwan), 73% of sub-provincial cities, and 35% of prefecture-level administrative regions had launched government data openness platforms, bringing the total number of such platforms to 142. In terms of the number of effective data sets opened through these platforms, 2020 saw a more than tenfold increase compared to 2017, covering key areas in each province, major departments, and pandemic prevention and control information (Fudan University Digital and Mobile Governance Laboratory, 2021).

Currently, some cities in China have already pioneered the sharing, circulation, and utilization of urban data. The National Shanxi Transformation Comprehensive Reform Demonstration Zone Planning Center has initiated the construction of a "data integration, sharing, and utilization-driven smart park informatization project construction and operation management system": firstly, it links data registration with project construction acceptance to ensure that all business system data meeting standards is promptly registered in the open database. Secondly, it establishes a data-sharing approval mechanism, enabling unified sharing of approved city data. Thirdly, it creates a data-sharing quality evaluation mechanism at the project level, whereby each encapsulated data set is evaluated by experts to promote the quality of shared data. Fourthly, it introduces a normalized data demand mining mechanism to continually enhance data mining capabilities for enterprises within the industrial park. As of January 2021, the project has successfully reduced duplicate data collection across over 150 new business systems, instead providing existing data through a shared open method to newly developed systems. Data from 16 business systems has been fully opened, consistently supplying high-quality urban data meeting data-sharing output standards. More than 500 types of urban data have been made available to park enterprises. Relying on urban data management and open utilization to benefit enterprises, the speed of enterprise transformation and product innovation has significantly increased. Within one year, the efficiency of park investment services and economic output both grew by over 20%.

In terms of international data openness practices, the UK government has made notable progress with its open data initiatives. As the world's largest government open data website, the UK government established the world's first Open Data Institute (ODI) in December 2012, with government funding. The ODI gathers world-class experts to assist enterprises in reforming R&D and exploring open data opportunities. It fosters innovation through collaboration, nurtures and guides new data-driven ideas, and by 2014 had 46 member organizations and 117,000 participants. Its goal is to collaborate with businesses to analyze UK government open data and cultivate innovative products and services (Qingbin, 2014). According to Yiu (2012), the UK government has saved an estimated £1 billion to £3 billion by reducing fraud and errors in the welfare system through open data. It has saved an additional £2 billion to £8 billion by effectively tracking tax evasion and avoidance. Regarding population censuses, the UK's Office for National Statistics (ONS) can assess the total UK population through open government data, enabling virtual censuses and eliminating the need for traditional census operations. This approach not only provides real-time census results but also saves the government over £500 million.

3 Challenges Facing Future Urban Governance

3.1 Risks of Applying Artificial Intelligence to Urban Governance

3.1.1 Security Risks of Intelligent Technology

Firstly, there is the issue of the vulnerability of technological systems. On one hand, issues such as power outages can render technology unusable. For example, in 2021, Henan Province experienced severe flooding caused by extreme rainfall, leading to widespread power and network outages, making even the most basic functions like mobile payments and telephone communication unavailable. If similar incidents occur in a scenario where artificial intelligence (AI) is widely applied, the negative impact on city operations and residents' lives would be immeasurable. On the other hand, the rapid development of AI has changed the risk points and attack methods in cybersecurity. Network attackers can exploit vulnerabilities to control AI algorithms, severely threatening the stability of urban operations and public safety. In the AI era, hackers are shifting their targets to critical infrastructure such as energy and transportation, which are closely tied to urban life. Especially in the future, when cities achieve "Internet of Everything" and "Intelligence of Everything", hacker intrusions could trigger a series of cascading effects.

Secondly, ethical safety is a major concern in AI development. As AI advances, people worry that it may overstep the behavioral boundaries set by ethical norms, posing threats to human existence. For instance, in 2017, two AI robots in Facebook's laboratory developed a machine language that only they could understand, but humans could not (Xinhua News, 2017). This case demonstrates that AI, through deep learning, may evolve "self-awareness" and might even replace humans in analysis and decision-making in the future. Based on current AI development trends, experts predict that machines with human-level intelligence could emerge by the mid-twenty-first century. Moreover, "superintelligent organizations" capable of mutual learning, interaction, and self-improvement may arise, potentially reversing roles to control and govern humans, posing a backlash risk (Deng, 2020).

Lastly, AI poses significant challenges to privacy and security. The widespread deployment of sensors, cameras, and other facilities across cities forms the foundation for smart urban governance. These devices continuously collect dynamic data on individual behaviors and social operations. During data collection, storage, and usage, issues such as data leaks, misuse, and illegal transactions may occur, severely compromising personal privacy, corporate interests, and national security. For example, in March 2018, Cambridge Analytica was exposed for improperly analyzing data from 50 million Facebook users, influencing the 2016 U.S. presidential election (Jiemian News, 2018). Such incidents have raised significant concerns globally. For instance, on May 16, 2019, the San Francisco government passed the Stop Secret Surveillance Ordinance, explicitly banning government agencies from using facial recognition technology and making it illegal to obtain information through such technologies (Observer Network, 2019).

3.1.2 Issues of Algorithmic Bias and Discrimination

In urban governance and public policy formulation, significant biases in AI algorithm predictions can trigger considerable political and social risks. When AI algorithms are based on biased data, the resulting policy recommendations are inherently flawed or only applicable to specific contexts, lacking universality (Wu et al., 2021).

The primary causes of algorithmic bias include inherent "bias" in data or insufficient data sample size. For instance, in 2015, Google Photos mistakenly labeled photos of two Black individuals as gorillas because the algorithm lacked sufficient training on dark-skinned image samples (Chuangjian.com, 2015). Similarly, in a beauty pageant judged by AI, most winners were white participants, as the training data primarily included images of white individuals. In urban settings, although much of the data is collected via devices, the placement of these devices may be human-determined, introducing inherent biases in data collection. Additionally, the "algorithmic black box" phenomenon contributes to bias (Liang, 2020). When deep learning algorithms become overly complex, it becomes challenging to trace the reasoning behind algorithmic decisions, making it difficult to identify and mitigate biased functions within the algorithm.

Algorithmic bias can amplify human biases. AI algorithms fundamentally use past experiences to predict future outcomes, but the discrimination and biases embedded in historical experiences may become entrenched in algorithms and amplified in future applications (Liang, 2020). When people tend to trust the conclusions derived from precise AI calculations, assuming that software is unbiased, they overlook the fact that these conclusions inherit biases from human value systems. Decisions based on these biased AI-driven conclusions may deviate from original intentions, further creating and reinforcing biased data for the algorithms. Overall, the biases exhibited by AI often originate from inheriting human value orientations and behavioral preferences during language learning. These biases, when reflected in algorithms, evolve into algorithmic discrimination. To address algorithmic bias and its resulting discrimination, it is essential to start by correcting human value orientations and behavioral preferences to "repair" these biases.

Furthermore, there is a need to guard against the potential "filter bubble" effect caused by algorithms. The term "filter bubble", coined by American scholar Cass Sunstein, refers to a phenomenon where users selectively consume information that aligns with their interests and preferences, forming a "personal newspaper" style information system. This narrows social cognition across different groups, leading to the rejection or ignorance of other perspectives and content (MBA Think Tank, 2021). In recent years, concerns about algorithms leading to "filter bubbles" have gained traction. The People's Daily has commented on algorithmic recommendation systems three times, highlighting that algorithms automatically filter out "uninteresting" or "disagreeable" information. Over time, this narrows public information reception channels, limiting diversity in information access (Yi, 2018). From a societal development perspective, filter bubbles may result in value polarization, conflict, and extremism among different groups, contributing to social instability.

3.1.3 The Conflict Between Technological Rationality and Humanistic Warmth

As artificial intelligence (AI) technology becomes more widely and deeply integrated, the conflict between technological rationality and humanistic warmth becomes increasingly prominent. The dominance of a technology-centric mindset is not only reflected in urban managers' blind pursuit, over-reliance, and improper use of technology but also in technology's "domination" over the essence and core of governance. Technology may lead to the "objectification" of governance, where decisions are made solely based on data and algorithms, lacking the necessary humanistic warmth. In the long term, the extensive use of technology could hinder urban managers' ability to innovate and think critically, stripping them of the capacity for rational, comprehensive, and dialectical analysis of urban issues. Simplifying complex urban governance problems into strings of code or data could render governance models rigid and indifferent.

Therefore, decisions cannot rely solely on data and technology while ignoring human initiative. Although AI technology can mimic human thinking and decision-making through deep learning, technology cannot balance rationality with emotional cognition, nor can it simultaneously address ethics, values, emotions, and culture while pursuing efficiency (Su et al., 2019). For urban managers, the core of governance should always be "people". When applying AI technology, they should avoid developing an absolute dependency on data and algorithms. Instead, algorithmic conclusions should serve as scientific foundations for governance or service decisions, enabling humans to exercise their initiative and make decisions imbued with humanistic warmth. Moreover, there should be ongoing reflection on the domination of technology and its dangers to mitigate potential negative impacts (Fu, 2021).

3.1.4 The Impact on Employment

AI has a dual effect on employment. On the one hand, AI can replace human labor in repetitive or unsafe jobs, enhancing labor productivity, addressing labor shortages in certain fields, and fostering new industries and business models. On the other hand, as robots become more efficient and their costs gradually fall below those of human labor, replacing humans with machines is an inevitable trend in the AI era. Unlike previous instances of mechanical replacements for physical labor, AI technology extends to many forms of intellectual labor, significantly expanding the scope of jobs that can be replaced (Agrawal et al., 2018).

The displacement effect of AI technology on low-skilled workers is indisputable. Frey and Osborne (2013) predicted that most jobs in transportation, passenger services, and public relations could be replaced by AI, with the probability of automation for real estate agents reaching 86%, sales and service personnel at 92%, and accounting and auditing at a staggering 94%. In China, many traditional labor-intensive industries in the secondary and tertiary sectors have become hotspots for unemployment under the trend of machine substitution (Jiang et al., 2018; Zhang, 2019; Zhou, 2017). In such circumstances, social inequality could worsen further. Specifically, some workers may accept lower wages to complete tasks far beyond the previous scope and volume of work simply to avoid being replaced by technology and falling into unemployment. This exacerbates the wealth gap, posing significant negative implications for social stability.

Intelligent technologies also exhibit a certain degree of substitution for high-skilled, high-income groups. A 2019 report by the Brookings Institution revealed that workers with higher education levels and incomes are increasingly at risk of job displacement. This is because some high-paying jobs (e.g., laboratory technicians and researchers) do not require the interpersonal skills that AI lacks, making them more susceptible to replacement by AI (Brookings Institution, 2019). A study jointly published by the Boston Consulting Group and the China Development Research Foundation also indicated that by 2017, 2.3 million jobs in the financial industry were eliminated due to the application of AI, accounting for about one-fourth of the workforce in the financial sector (Boston Consulting Group et al., 2018). Evidently, the profound impact of AI on employment structures, even in high-end sectors such as finance, cannot be overlooked.

3.2 *Challenges Brought by Data Openness*

3.2.1 Incomplete Legal and Regulatory Framework for Data Openness and Security

The Data Security Law of the People's Republic of China, passed on June 10, 2021, is the country's first foundational legislation in the field of data security. It explicitly regulates government data openness and reflects the legislative goal of a holistic national security perspective. The law focuses on prominent issues in the field of data security, establishing systems for data classification and hierarchical management, risk assessment, monitoring and early warning, emergency response, and data security review, while also clarifying the data security protection obligations of relevant entities (Wang et al., 2021). However, the development of China's legal framework for data openness and security still lags behind the actual practices of government data openness.

First is the lack of government data open licensing agreements. Government data openness requires not only national laws and policy guarantees but also binding licensing agreements. Countries like the United Kingdom (UK Government Licensing Framework), the United States (CC0 Public Domain Dedication), and

France (Licence Ouverte) have established government data open licensing agreements that effectively protect individual privacy and national security. However, China currently lacks any licensing agreement to support the deep openness and utilization of government data (Huang & Miao, 2017).

Second is the insufficient local exploration. Local legislative efforts for government data sharing and openness are limited to a few cities, such as The Guiyang Municipal Regulations on Government Data Sharing and Openness and The Interim Measures for Public Data Openness in Shanghai. Yet much of the data is concentrated at the municipal or grassroots level, mainly for local government and residents. Therefore, cities must establish localized legal and regulatory systems tailored to practical local needs, based on national laws and regulations.

3.2.2 Fragmentation of Data Resources

Firstly, many departments have developed excessive systems and databases purely for the sake of informatization. The institutional and organizational fragmentation in government also results in data silos between departments. For example, during the "Golden Week" holidays, severe traffic congestion occurs in many major cities. One major reason is that transportation and cultural tourism systems lack data interoperability. Traffic departments cannot access real-time data from cultural and tourism departments, making it difficult to predict and address potential congestion.

Secondly, the lack of open and interoperable data between cities hinders population mobility and the equalization of public services. Take health codes as an example. While the introduction of health codes in March 2020 greatly facilitated epidemic prevention and economic recovery, nearly a hundred different health code systems were developed across provinces and cities. The lack of data openness and mutual recognition between regions made it difficult for many workers to travel.

Lastly, cities often focus on opening data collected by various departments and levels, neglecting to release other data more closely related to citizens' lives, such as ecological environment monitoring data and city-wide IoT data. Furthermore, the integration of government data with multi-source social big data is insufficient. Possible reasons include the lack of mechanisms for businesses to share data with governments, a strong "data ownership" mindset among businesses, and insufficient openness of government data (Guo et al., 2019).

3.2.3 Shortage of Data Talent

Currently, global data volume is growing at an average annual rate of 50%. According to IDC consulting predictions, the total global data volume reached 44 ZB in 2020, with China accounting for over 8 ZB or 18% of the global total. By 2025, the global data volume is expected to grow to 175 ZB (SD Technology Manufacturing, 2021). This massive volume of data also signals a rising demand for data scientists and data professionals. Additionally, in urban management, there is a strong demand for professionals proficient in both digital technology and urban operations across various departments.

The shortage of data talent is a common issue in cities across China and globally. Internationally, in 2014, The Wall Street Journal noted that the profession of data scientist barely existed before 2013, but it has since become one of the hottest jobs in the high-tech labor market (Wu, 2014). A 2011 report by the McKinsey Institute also predicted that by 2018, the United States could face a shortage of 1.5 million professionals in big data-related management and analysis roles (Tu, 2014). Domestically, as early as 2010, experts had projected that China would need at least 1 million big data professionals over the following five years, but the country had fewer than 100,000 at the time (Chen, 2015). According to the Digital Economy Employment Impact Report released by the China Academy of Information and Communications Technology, China faced a shortage of nearly 11 million digital talents in 2020, and this gap continues to widen (Guangming Online, 2021). Evidently, data talent has become a critical human resource for enhancing the core competitiveness of nations and major cities in the digital era.

In summary, while the development of artificial intelligence and the trend toward data openness bring numerous opportunities for future urban governance, they also present many uncertain challenges that require preparation to address. When widely applying artificial intelligence technologies, potential risks such as security vulnerabilities, algorithm bias, and algorithmic discrimination must be considered. Additionally, the results derived from technology and data calculations must be viewed rationally, and final decisions should reflect a strong sense of humanistic care. Regarding the risks that artificial intelligence applications may pose to employment, governments must implement a range of policies and measures to mitigate the negative impact on jobs and maintain social stability and harmony. Amid the continued growth of open databases, issues such as the incomplete legal framework for data openness and security, data resource fragmentation, and the shortage of data talent must also receive adequate attention and resolution.

References

Agrawal, A., Gans, J., & Goldfarb, A. (2018). *Prediction machines: The simple economics of artificial intelligence*. Harvard Business Press.

AI Chip World. (2021). *2020 White paper on knowledge graph industry for AI new infrastructure*. Retrieved December 31, 2021, from https://www.163.com/dy/article/FTOLOFSG05383ZUN.html

Ai, L., & Wang, G. (2015). *Reshaping public-oriented government services*. Social Sciences Academic Press.

Alibaba DAMO Academy. (2020). *Top 10 technology trends in 2020*. Retrieved December 31, 2021, from https://damo.alibaba.com/events/57

Boston Consulting Group & China Development Research Foundation. (2018). *Replacement or liberation: The impact of AI on the financial labor market*. Retrieved December 31, 2021, from https://web-assets.bcg.com/img-src/BCG-CDRF-The-Impact-of-AI-on-the-Financial-Job-Market_Mar%202018_CHN_tcm9-187844.pdf

Brookings Institution. (2019). *Automation and artificial intelligence: How machines are affecting people and places*. Retrieved December 31, 2021, from https://www.brookings.edu/research/automation-and-artificial-intelligence-how-machines-affect-people-and-places/
CCTV News. (2021). Shanghai data exchange officially launched today. https://china.huanqiu.com/article/45j4VmdhKbK.
Chen, G. L. (2015). Strengthening the construction of big data talent teams. *China Construction Information, 7*, 21.
Chen, S. L. (2016). Theoretical basis of government data openness. *Theory and Reform, 6*, 104–107.
Chen, T., Ran, L. Y., & Ming, C. H. (2018). Research on the application of artificial intelligence in government services. *E-Government, 3*, 22–30.
China Jiangsu Network. (2021). Wuxi builds a comprehensive protection system, creating a protective net for urban cybersecurity. https://china.huanqiu.com/article/45D3IwLZFYu.
China News Network. (2012). California, USA, allows driverless cars to legally operate on public roads. https://www.chinanews.com.cn/gj/2012/09-26/4213051.shtml.
Chuangjian.com. (2015). *Google Photos app mistakenly labeled Black individuals as gorillas*. Retrieved December 31, 2021, from https://www.chinaz.com/news/2015/0702/418974.shtml
Chongqing Daily. (2021). Sichuan and Chongqing sign 1+9 big data collaborative development cooperation agreement. http://www.cq.gov.cn/ywdt/jrzq/202108/t20210828_9632806.html.
Deng, W. Q. (2020). Risks and institutional prevention in the era of artificial intelligence. *People's Forum, 34*, 69–71.
Dong, K. N., & Wang, N. (2017). The dawn of intelligent healthcare: An overview of AI + health applications. *Big Data times, 4*, 26–37.
Du, Z. (2017). Intelligent robots make elderly care more accessible. *China Strategic Emerging Industries* (13), 48–50. https://doi.org/10.19474/j.cnki.10-1156/f.000882
Feng, J., & Qi, Y. P. (2021). Application of knowledge graphs in smart urban governance. *China National Conditions and Strength* (7), 50–53. https://doi.org/10.13561/j.cnki.zggqgl.2021.07.014
Frey, C. B., & Osborne, M. (2013). *The future of employment*.
Fudan University Digital and Mobile Governance Lab. (2021). *China local government data openness report (Second Half of 2020)*. Retrieved December 31, 2021, from http://dsj.guizhou.gov.cn/xwzx/gnyw/202101/t20210122_66269368.html
Fu, X. R. (2021). Major risks of smart governance and their mitigation strategies. *National Governance, 14*, 31–34.
Fu, X. W., & Zheng, L. (2013). A review of domestic research on government data openness. *E-Government, 6*, 8–15.
Gao, Q. Q., & Liu, Y. (2019). Urban governance in the era of artificial intelligence. *Journal of Shanghai Administration Institute, 2*, 33–42.
Gao, Q. Q., & Lü, J. Y. (2017). Smart healthcare: Opportunities and challenges for public health in the AI era. *E-Government, 11*, 11–19.
Guangming Online. (2021). *Digital economy employment impact research report released: Great potential for flexible employment on digital platforms*. Retrieved December 31, 2021, from http://news.china.com.cn/mts/2021-09/30/content_1174154.htm
Guo, M. J., Tong, N. N., & Wang, J. D. (2019). Problems and countermeasures in the sharing and utilization of government and social data. *China Economic & Trade Herald, 8*, 37–38.
Hengxing IoT. (2021). *AI + IoT: From the Internet of Things to the Internet of Intelligence*. Retrieved December 31, 2021, from https://zhuanlan.zhihu.com/p/94150658
Huang, R. H., & Miao, M. (2017). Security protection strategies for open government data in China. *E-Government, 5*, 28–36.
Jiang, N. P., & Zou, Y. (2018). Artificial intelligence and China's labor supply-side structural reform. *Journal of Sichuan University: Philosophy and Social Sciences Edition, 1*, 130–138.
Jiangmen Radio and Television Station. (2021). A visual guide to the Jiangmen-Hong Kong Cross-Border Government Service Zone. https://wap.peopleapp.com/article/rmh23744819/rmh23744819.

Jiemian News. (2018). Cambridge Analytica, embroiled in Facebook data scandal, declares bankruptcy. https://baijiahao.baidu.com/s?id=1599499596676990548&wfr=spider&for=pc.
Jin, G. Q. (2019). Application of knowledge graph technology in smart cities. *China Surveying and Mapping, 10*, 62–64.
Li, Y. D. (2008). A preliminary study on the application of artificial intelligence in E-Government construction. *E-Government, 5*, 87–93.
Li, J. Z., & Hou, L. (2017). A review of knowledge graph research. *Journal of Shanxi University (Natural Science Edition), 40*(3), 454–459.
Li, Z., Zhou, D. H., & Wang, Y. (2019). Educational knowledge graph under the perspective of Artificial Intelligence+: Connotation, technical framework and application research. *Journal of Distance Education, 37*(4), 42–53.
Liberation Daily. (2016). Government information disclosure: Shanghai ranks first in provincial government transparency. https://china.huanqiu.com/article/9CaKrnJUFcw.
Lin, L. P. (2018). Research on the application of artificial intelligence in e-government public services. *Automation Applications, 7*, 157–158.
Liu, D. J., Du, J., Jiang, N., & Huang, R. H. (2018). The development trend of artificial intelligence integration into school eEducation. *Open Education Research, 24*(4), 33–42.
Liang, X. F. (2020). Reflections on algorithmic discrimination in the era of artificial intelligence. *China Informationization, 7*, 54–55.
Internet Medical and Healthcare Industry Alliance. (2018). *White paper on medical AI technology and application*. Retrieved December 31, 2021, from https://www.sohu.com/a/230653403_483389
Ma, H. Q., & Pu, P. (2016). An exploration of the connotation and theoretical basis of open data. *Library Theory and Practice, 11*, 48–54.
MBA Think Tank. (2021). *Information cocoons*. Retrieved December 31, 2021, from https://wiki.mbalib.com/wiki/Information_Cocoons
Observer Network. (2019). *San Francisco votes to ban police from using facial recognition technology; opponents fear crime rates may rise*. Retrieved December 31, 2021, from https://www.guancha.cn/internation/2019_05_15_501681.shtml
Qing, B. (2014). Seizing data opportunities: The UK's data capability strategy. *Information Theory and Practice, 37*(5), 145.
Qiu, W. W. (2017). Application and challenges of artificial intelligence technology in government services. *Electronic Technology & Software Engineering, 20*, 259–260.
SD Technology Manufacturing. (2021). *IDC releases the latest big data spending guide: China's total big data market will exceed $25 billion by 2025*. Retrieved December 31, 2021, from https://www.sohu.com/a/484571985_121124366
Shenzhen Special Zone Daily. (2020). Shenzhen takes the lead in local data legislation, releasing a draft for public comment. http://sztqb.sznews.com/PC/content/202007/17/content_889852.html.
Snowball. (2020). How to invest in ETFs during a period of transition. Artificial intelligence is ubiquitous in our lives – Autonomous driving. https://xueqiu.com/6158882456/158660047.
Song, H. L. (2018). The development and construction of China's legal system for open data. *Administrative Law Research, 2*, 35–46.
Su, J. & Huang, C. (2019). Exploring Chinese solutions for social governance of artificial intelligence. https://epaper.gmw.cn/gmrb/html/2019-12/26/nw.D110000gmrb_20191226_1-16.htm.
Su Lun AI Channel (2020). AI report. From perceptual intelligence to cognitive intelligence. https://zhuanlan.zhihu.com/p/124959650.
Suzhou Daily. (2020). Adaptive traffic lights talk, safeguarding children's safety. http://js.news.163.com/20/0720/10/FHVKIU3604248E8R.html.
Tang, S. S., & Liu, Y. T. (2018). Strengthening Data Governance to Enhance National Innovation Capacity. http://www.sic.gov.cn/News/612/9707.htm.
Tu, T. (2014). "Data Scientist" is becoming the most popular profession. Science and Technology Daily, 2014-06-13(005).
Wang, X. D. (2021). Problems and countermeasures facing the opening up of public data in my country. *China Economic and Trade Guide, 10*, 78–79.
Wang, Z. G. (2017). Robots enter the era of intelligent elderly care. *Smart China, 1*, 86–87.

Wang, C. H., & Cheng, L. (2021). Analysis of ten key legal points of the data security law of the People's Republic of China. https://m.gmw.cn/baijia/2021-09/01/35130741.html.

Wu, C. L. (2014). "Data Analysis" has become a "golden job". *People's Daily*, 2014-08-21(022).

Wu, J. J., & Fu, Y. (2021). Algorithmic decision-making: Public decision-making driven by artificial intelligence and its risks. *Open Era, 5*, 194–206.

Xinhua News Agency. (2017). Robots create their own language chat? Artificial intelligence: Should I love you or fear you?. http://www.xinhuanet.com/world/2017-08/03/c_129671008.htm.

Yang, Z., Yang, N., & Xu, M. J. (2016). Research on artificial intelligence-related technologies for Internet of Things applications. *Telecommunications Technology, 5*, 16–19.

Yi, Z. (2018). "The Screaming Effect" and "Information Cocoon". http://m.people.cn/n4/2018/0502/c25-10910831.html.

Yiu, C. (2012). *The big data opportunity*. Policy Exchange.

Zhang, Y. X. (2018). Urban governance under the digital wave: From refinement to intelligence. *Telecommunications Network Technology, 3*, 9–11.

Zhang, Y. Z. (2019). The employment effects of artificial intelligence and robots and countermeasures. *Scientific Management Research, 37*(1), 43–45.

Zhang, Y. N. (2020). On the path and choice of legislation for open government data—taking the Guiyang Municipal Government Data Sharing and Opening Regulations as an example. *Chinese Public Administration, 2*, 38–44.

Zhang, Q. F., & Lv, Y. B. (2002). On the right to know. *Journal of Jiangsu Academy of Administration, 1*, 106–114.

Zhang, K. Y., & Zhang, J. N. (2017). New areas, misconceptions, blind spots and forbidden zones in the application and research of artificial intelligence in education. *Journal of Distance Education, 35*(5), 54–63.

Zhao, F., Lan, L., Cao, Z. Q., Sun, H., Yin, X., & Jin, Z. Y. (2018). Research on the current status of artificial intelligence application and development in the field of health care in my country. *China Health Information Management Journal, 15*(3), 344–349.

Zhou, W. B. (2017). Research on the impact of robot applications on human resource management. *Journal of Nanjing University: Philosophy, Humanities and Social Sciences, 54*(6), 23–24.

Zhejiang Provincial "Run Once at Most" Reform. (2020). Wenzhou's Personal Data Treasure releases public data convenience benefits. https://zj.zjol.com.cn/red_boat.html?id=100851740.

MIX
Papier aus verantwortungsvollen Quellen
Paper from responsible sources
FSC® C105338

If you have any concerns about our products,
you can contact us on
ProductSafety@springernature.com

In case Publisher is established outside the EU,
the EU authorized representative is:
Springer Nature Customer Service Center GmbH
Europaplatz 3, 69115 Heidelberg, Germany

Printed by Libri Plureos GmbH
in Hamburg, Germany